Event Horizon Strategy

Event Horizon Strategy

Enhance Careers, Create Business Value,
and Avoid Industry Disruption with
Generative AI

Jason Davis

Dragon Dance Publishing

Jason Davis

Copyright © 2024 All rights reserved.

Event Horizon Strategy

Enhance Careers, Create Business Value, and Avoid Industry Disruption with Generative AI

Table of Contents

Table of Contents

Table of Contents

Chapter 1

—

The Strategic Challenge of Generative AI

"Everyone is now talking about AI, but few have the faintest glimmer of what is about to hit them." – Leopold Aschenbrenner[1]

"We're building progressively greater intelligence. The percentage of intelligence that is not human is increasing. And eventually we will represent a small percentage of intelligence." – Elon Musk[2]

"The future is going to be good for the AIs regardless; it would be nice if it would be good for humans as well." – Ilya Sutskever[3]

---

[1] Aschenbrenner, L. (2024). Situational Awareness: The Decade Ahead. https://situational-awareness.ai/wp-content/uploads/2024/06/situationalawareness.pdf

[2] Musk, E. (2020). Elon Musk Podcast Transcript, *Joe Rogan Experience.* May 7, 2020

[3] Heaven, W. D. (2023). Rogue superintelligence and merging with machines: Inside the mind of OpenAI's chief scientist. *MIT Technology Review.* October 26, 2023

Generative AI is reshaping how people, businesses, and societies interact with and use technology, thanks to the rapid evolution of Large Language Models (LLMs). These models are excellent not only at encoding linguistic patterns, but also at producing coherent, creative results. Early innovations, such as DALL-E and Stable Diffusion, expanded AI's creative capabilities by transforming text into vivid, lifelike images. However, OpenAI's ChatGPT had a widespread impact, redefining interactive AI with unprecedented user adoption. In response, major players such as Microsoft and Google have launched comparable LLMs, along with a growing number of open-source alternatives and emerging startup innovations in generative AI.

LLMs can be seen as the next significant leap forward in augmented human intelligence. The current generation of LLMs encode virtually all human knowledge and wisdom that is publicly available on the internet, and sometimes proprietary data, making it easily accessible. But LLM capabilities go beyond simply encoding knowledge and make it possible to generate new knowledge based on inference from its training set.[4] As venture capitalist Marc Andreessen recently noted, this intelligence amplification impact is so general that it could impact virtually every area of human endeavor, from work to art to governance.[5] This significant increase in intelligence offers vast potential to solve problems and improve lives, positioning

---

[4] Jia, N., Luo, X., Fang, Z., & Liao, C. (2024). When and how artificial intelligence augments employee creativity. *Academy of Management Journal*.

[5] Andreessen, M. (2023a). *An even shorter description of what AI could be: A way to make everything we care about better.* . 2023-06-06 https://twitter.com/pmarca/status/1666112508426608640

generative AI as a quantum leap in intelligence with wide-ranging applications.

The rapid adoption and immense demand speak for themselves. The number of LLM users reached 100 million users in a few months, one of the fastest product adoption waves in history.[6] OpenAI reported in August 2024 that more than 200 million people use ChatGPT every week, suggesting that its number of users doubled in less than a year.[7] AI adoption related to machine learning was already accelerating in the United States,[8] but the emergence of LLMs represents a qualitative change. This explosive growth outpaces that of even the most popular social media platforms like TikTok, Instagram, and Facebook, as well as streaming giants like Netflix and Spotify, despite generative AI being largely text-based.

What is driving the adoption of generative AI? Its potential to considerably impact work, jobs, and careers is undoubtedly one of the drivers of individual adoption. Some early examples suggest that the nature of work is being reshaped by generative AI. For instance, graphic design is evolving as generating images from text has become more accessible, marketers are creating advertisements and web content with generative AI solutions, and patients are consulting doctors about treatments suggested by ChatGPT.

---

[6] Hu, K. (2023). *ChatGPT sets record for fastest-growing user base.* Reuters. Feb 2, 2023 https://www.reuters.com/technology/chatgpt-sets-record-fastest-growing-user-base-analyst-note-2023-02-01/

[7] Roth, E. (2024). ChatGPT's weekly users have doubled in less than a year / Now 200 million people use the AI chatbot each week. *The Verge.* April 30, 2024 https://www.theverge.com/2024/8/29/24231685/openai-chatgpt-200-million-weekly-users

[8] McElheran, K., Li, J. F., Brynjolfsson, E., Kroff, Z., Dinlersoz, E., Foster, L., & Zolas, N. (2024). AI adoption in America: Who, what, and where. *Journal of Economics & Management Strategy, 33*(2), 375-415.

In fact, benchmarking AI systems against average human performance in prototypical tasks has become an important test of these systems' capabilities in various spheres. A recent meta-analysis indicated that AI systems have already exceeded human performance across a wide variety of tasks, although not necessarily every task.[9] Tasks where generative AI excels include handwriting recognition, language understanding, speech recognition, sentence completion, image recognition, grade school math, and reading comprehension.[10] A Stanford University AI report indicates that AI exceeds average human levels in visual reasoning, reading comprehension, mathematics, and language understanding.[11]

---

[9] Dell'Acqua, F., McFowland, E., Mollick, E. R., Lifshitz-Assaf, H., Kellogg, K., & Lakhani, K. R. (2023). *Navigating the jagged technological frontier: field experimental evidence of the effects of AI on knowledge worker productivity and quality* (Harvard Business School Technology & Operations Mgt. Unit Working Paper, Issue.

[10] Douwe, K. (2023). *Plotting Progress in AI*. Contextual AI. 2023-05-23 https://contextual.ai/news/plotting-progress-in-ai/

[11] Murad. (2024). *AI and Automation is going to be replacing more & more jobs every year (Stanford Study)*. 2024-02-22 https://x.com/MustStopMurad/status/1787795402168623130/

MAN VS. MACHINE

AI Models Are Improving Every Year

AI Technical Performance [Selected measures, 100% = human baseline]

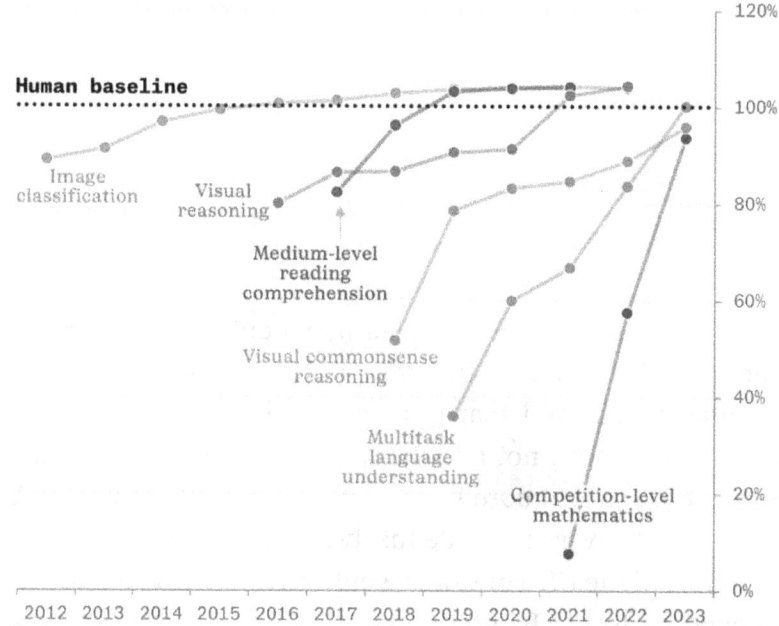

Human baseline

Image classification

Visual reasoning

Medium-level reading comprehension

Visual commonsense reasoning

Multitask language understanding

Competition-level mathematics

120%

100%

80%

60%

40%

20%

0%

2012 2013 2014 2015 2016 2017 2018 2019 2020 2021 2022 2023

CHARTR

Source: Stanford University AI Index Report 2024

Other data indicate similar improvements in capabilities.[12] That is, AI appears to be exceeding benchmarks not only in professional domains, but in terms of more basic human capabilities as well.

---

[12] Douwe, K. (2023). *Plotting Progress in AI.* Contextual AI. 2023-05-23 https://contextual.ai/news/plotting-progress-in-ai/

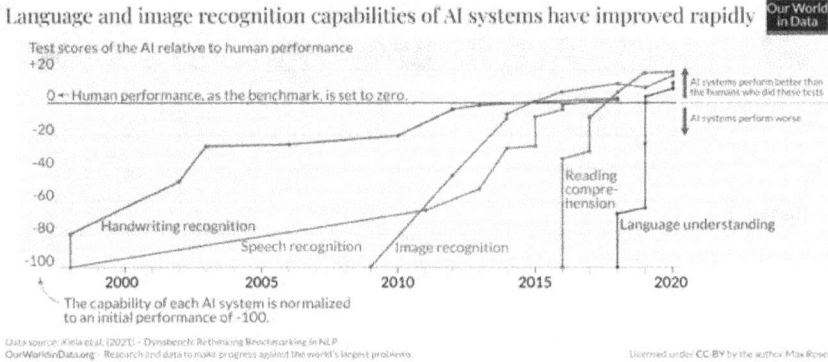

Language and image recognition capabilities of AI systems have improved rapidly

The capability of each AI system is normalized to an initial performance of -100.

The implications for employment are enormous. Goldman Sachs estimates that 7% of US employment could be substituted by AI, and that 300 million jobs could be at risk globally.[13] Of course, not all jobs may be affected the same, as some work tasks are more substitutable than others by AI. An OpenAI and University of Pennsylvania study estimates that around 80% of the US workforce would have at least 10% of their tasks exposed to LLMs, potentially leading to a shift in career paths as the costs and benefits of learning tasks change.[14] Some scenarios foresee a dramatic reduction in wages even as productivity soars.[15] While currently unclear, the implications of AI on employment and work are likely to be enormous.

---

[13] Covello, J. (2024). *Gen AI: Too Much Spend, Too Little Benefit?* Goldman Sachs. https://www.goldmansachs.com/images/migrated/insights/pages/gs-research/gen-ai--too-much-spend,-too-little-benefit-/TOM_AI%202.0_ForRedaction.pdf

[14] Eloundou, T., Manning, S., Mishkin, P., & Rock, D. (2023). Gpts are gpts: An early look at the labor market impact potential of large language models. *Working Paper*. https://arxiv.org/abs/2303.10130. , OpenAi. (2023). *GPT-4 Technical Report.* https://cdn.openai.com/papers/gpt-4.pdf

[15] Korinek, A. (2023b). Scenario Planning for an A(G)I Future. *International Monetary Fund.*

What drives such high demand for generative AI, and what is it that drives its considerable impact? Generative AI outpaces existing AI and other technologies, showing a truly exponential and non-cyclical growth pattern. Its ability to recall and reproduce useful knowledge is a key to its value. This is demonstrated in a variety of tests, from bar exams to SATs, and in various educational contexts.[16] LLMs, trained on nearly all publicly available content on the internet, can generate a wide array of content based on this massive corpus of knowledge.[17]

https://www.imf.org/en/Publications/fandd/issues/2023/12/Scenario-Planning-for-an-AGI-future-Anton-korinek

[16] Douwe, K. (2023). *Plotting Progress in AI*. Contextual AI. 2023-05-23 https://contextual.ai/news/plotting-progress-in-ai/ -

[17] Aschenbrenner, L. (2024). Situational Awareness: The Decade Ahead. https://situational-awareness.ai/wp-content/uploads/2024/06/situationalawareness.pdf

Performance on common exams
(percentile compared to human test-takers)

	GPT-4 (2023)	GPT-3.5 (2022)
Uniform Bar Exam	90th	10th
LSAT	88th	40th
SAT	97th	87th
GRE (Verbal)	99th	63rd
GRE (Quantitative)	80th	25th
US Biology Olympiad	99th	32nd
AP Calculus BC	51st	3rd
AP Chemistry	80th	34th
AP Macroeconomics	92nd	40th
AP Statistics	92nd	51st

SITUATIONAL AWARENESS | Leopold Aschenbrenner

Test scores of AI systems on various capabilities relative to human performance

Our World in Data

Within each domain, the initial performance of the AI is set to -100. Human performance is used as a baseline, set to zero. When the AI's performance crosses the zero line, it scored more points than humans.

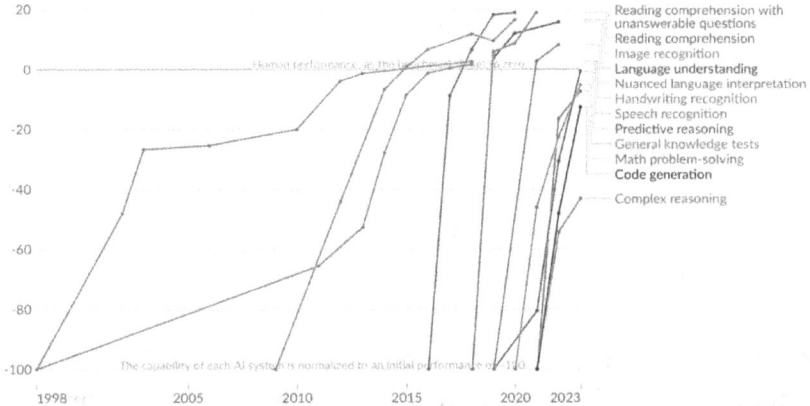

Data source: Kiela et al. (2023)   OurWorldInData.org/artificial-intelligence | CC BY
Note: For each capability, the first year always shows a baseline of -100, even if better performance was recorded later that year.

Yet, it is the capacity for applying and generating knowledge that may have the biggest impact on productivity during work. MIT researchers estimate a 55.8% improvement in productivity for people who use generative AI, whereas an experimental study found a significant productivity boost for programmers who use AI-based tools like as Microsoft's GitHub Copilot.[18] More broadly, Goldman Sachs predicts generative AI could raise US productivity growth by 1.5% over 10 years.[19]

---

[18] Peng, S., Kalliamvakou, E., Cihon, P., & Demirer, M. (2023). The Impact of AI on Developer Productivity: Evidence from Github Copilot. *Working Paper*. https://arxiv.org/abs/2302.06590

[19] Lee, G. (2024). *AI could increase growth by 1.5% over the next 10 years, Goldman Sachs says*. CNBC. February 14, 2024 https://www.cnbc.com/video/2024/02/14/ai-could-increase-growth-by-1point5percent-over-the-next-10-years-goldman-sachs.html

Given the enormous demand for generative AI and its significant impact, it is unsurprising that both individuals and businesses are showing strong interest. This is visible in internet search data, for example, where searches for AI use cases have increased dramatically. For example, shortly after its launch, more Google users looked up ChatGPT than any other AI platform. However, some research show that workers are also concerned about job replacement. According to Google Trends statistics, "Will AI replace me?" became one of the most sought phrases shortly after ChatGPT was introduced.

Company executives are particularly keen on exploring generative AI. A 2023 KPMG survey shows that 65% believe generative AI will have a high or extremely high impact on their organization in the next 3-5 years.[20] Additionally, 60% say they are 1-2 years away from implementing their first generative AI solution. Executives are most optimistic about opportunities to increase productivity (72%), change the way people work (65%), and encourage innovation (66%). Indeed, this may even be accelerating. A Wharton study of 800 senior managers found that usage has doubled in 2024, with the majority reporting positive impact.[21]

The earliest window into the impact of generative AI on organizations is in the field of software development. Big tech companies with thousands of developers have been the earliest and most intense adopters of generative AI systems, perhaps

---

[20] KPMG. (2023). *KPMG U.S. survey: Executives expect generative AI to have enormous impact on business, but unprepared for immediate adoption.* https://kpmg.com/us/en/media/news/kpmg-generative-ai-2023.html

[21] Korst, J., Puntoni, S., & Purk, M. (2024). *Growing Up: Navigating Gen AI's Early Years.* AI at Wharton and GBK Collective. https://ai.wharton.upenn.edu/wp-content/uploads/2024/11/AI-Report_Full-Report.pdf

because software engineering is an area in which LLMs are a particularly effective tool. Amazon's CEO Andy Jassy spoke about the impact of Amazon Q, a GenAI system developed at Amazon to assist software development.[22] "The average time to upgrade an application to Java 17 plummeted from what's typically 50 developer-days to just a few hours," he noted, adding that "we estimate this has saved us the equivalent of 4,500 developer-years of work (yes, that number is crazy but, real)." The impact was not just on developer time, but also on actual improvements to the software and IT systems themselves. For instance, Jassy stated that "the upgrades have enhanced security and reduced infrastructure costs, providing an estimated $260M in annualized efficiency gains." We can expect more of these examples of organizational gains to emerge as time goes on.

The ultimate economic impact of generative AI will be substantial. A 2023 McKinsey study estimates generative AI could add an equivalent of $2.6 trillion to $4.4 trillion to the global economy annually through new use cases, and another $6.1 to $7.9 trillion due to increased worker productivity.[23] These gains will impact many organizational functions and industries, but value creation might be concentrated in a few key areas where content generation is crucial: customer operations, marketing and sales, software engineering, and R&D.

---

[22] Jassy, A. (2024). *Amazon Q, our GenAI assistant for Software Development.* 2024-08-22
https://x.com/ajassy/status/1826608791741493281
[23] Chui, M., Roberts, R., Yee, L., Hazan, E., Singla, A., Smaje, K., Sukharevsky, A., & Zemmel, R. (2023). The economic potential of generative AI. *McKinsey Report.* June 2023
https://www.mckinsey.com/capabilities/mckinsey-digital/our-insights/the-economic-potential-of-generative-ai-the-next-productivity-frontier

Providers of AI will capture a significant portion of this value creation. An Ark Investment study estimates that at 100% adoption, AI spend of $41 trillion could increase global labor productivity by $200 trillion by 2030. By the same year, AI software could generate $14 trillion in revenue and $90 trillion in enterprise value if vendors capture even 10% of the value created by their products.[24] Similarly, Bloomberg estimates that Generative AI will generate $1.3 trillion in revenue by 2032, making up 12% of all technology spend.[25]

The advent of generative AI presents an era of mass abundance, marked by job creation and faster, cheaper, and better production of goods and services. As we delve further in this book, we will explore the multifaceted implications of this transformative technology.

Generative AI is proving to be a game-changer in the entrepreneurial world too. At a time when startup funding was generally down compared to prior years, the influx of generative AI startups broke the trend. Investments in the sector have skyrocketed, with total investment in the sector reaching $21 billion in 2023 from $5 billion only a year earlier, according to Pitchbook.[26] Regardless of the precise figure, it is clear that generative AI has sparked a notable surge in entrepreneurial

---

[24] Management, A. I. (2023). *Ark Investment 2023 Generative AI.* Ark Investment Management LLC. January 31, 2023 Ark Investment Management LLC

[25] Bloomberg. (2024). *Generative AI 2024 Report.* https://www.bloomberg.com/professional/products/bloomberg-terminal/research/bloomberg-intelligence/download/generative-ai-2024-report/

[26] Robins, J. (2023). *Meet the generative AI startups pulling in the most cash.* Pitchbook. October 18, 2023 https://pitchbook.com/news/articles/vc-valuations-generative-ai-startups

activities—also visible in the mounting number of AI-centered hackathons, particularly in tech hotspots like San Francisco.

This isn't only the case in developed economies. A recent study of Kenyan entrepreneurs suggests, for example, that new companies in emerging markets—where access to technology and capital have traditionally been scarce—may be a particularly good place of application for generative AI, as evidenced by improved business performance among startups using it.[27] Generative AI is especially appealing and of inherent value to individual entrepreneurs and small-team startups as they can leverage open source tools to expedite and better their projects.

Even traditional firms, which are not often associated with cutting-edge technology, are joining the bandwagon. Many people are investigating the possibility of developing their own businesses and services using generative AI. Even those who are not starting new ventures are attempting to maximize the value of this technology by implementing AI technologies. These attempts span industries, from healthcare to journalism, as corporations see the value of AI-generated content.

On the flip side, generative AI offers a substantial danger to enterprises that are hesitant to implement it. If employed correctly, this technology can provide a competitive advantage, producing an imbalance in favor of those that adopt it first. In this competition, startups and large technology companies are currently outperforming existing enterprises that are resistant or slow to adapt. This might cause industry-wide shocks, changing market dynamics and leaving behind firms that are unable to keep up with the AI revolution.

---

[27] Otis, N. G., Clarke, R., Delecourt, S., Holtz, D., & Koning, R. (2023). The uneven impact of generative AI on entrepreneurial performance. *Working Paper*. https://www.hbs.edu/ris/Publication%20Files/24-042_9ebd2f26-e292-404c-b858-3e883f0e11c0.pdf

Despite the potential for disruption and adverse outcomes, the overall scenario is good in terms of value creation. The influence of generative AI is anticipated to spread over a wide range of individuals, businesses, and industries, redefining processes and productivity norms along the way. However, this technological boon is not devoid of disparities.

One potential outcome is that the benefits of generative AI are tightly concentrated. A small elite of entrepreneurs, employees, and organizations might be the ones making significant strides and capturing the majority of the value. The trend of adoption might follow an accelerated yet unequal pattern, with many holding off from integrating this technology into their workflows for as long as possible. The unequal distribution of benefits and rewards can have far-reaching social consequences. Robot taxes and Universal Basic Income (UBI) have been offered as methods to rectify this disparity and spread AI's value more fairly. A more useful and potentially influential approach would be to promote widespread technology use and ownership.

More vulnerable employees and organizations could benefit from generative AI, particularly if they can own some of the outputs, potentially minimizing the disruptive impact. Furthermore, promoting the creation of new businesses and applications may widen the benefits of AI, adding value to all sectors of society.

Generative AI's profound and rapid influence makes it a possible contender for the most momentous technological revolution of our time. Individuals, supervisors, and entrepreneurs should take note. While the potential benefits are substantial, careful attention is essential to guarantee that they are dispersed equitably and that the entire society can adapt to fully realize the potential of this exceptional technology.

Outline of the Book

The core challenge this book tackles is the inherent uncertainty that comes with the rapid evolution of generative AI. Despite its disruptive potential and multitude of opportunities, individuals and companies are struggling to anticipate its trajectory and identify the best ways to leverage the technology. By gaining insights into the early best practices employed by pioneering individuals and organizations, the adoption of generative AI can be expedited and its value can be fully harnessed for the benefit of all.

The book adopts a technology strategy approach, putting innovation and entrepreneurial action at its heart. In an area in which the future evolution of technology and its optimal use cases are uncertain, the most effective approach is to encourage action through experimentation. It subscribes to the principle of learning by doing—taking measured risks based on what is presently understood about the technology.

But where does one begin? This book provides guidance on that crucial question, backed by extensive and integrative research into understanding generative AI and the applications that have proven successful in creating value. It explores the scope of possibilities within generative AI, laying down frameworks and exemplary models to inspire and guide readers in applying this technology effectively.

Key Frameworks

One of the key contributions of this book is the Event Horizon Analysis Framework, that contrasts three strategies for dealing with generative AI. The book also highlights three Generative AI Product Architectures (Content Generators, Copilot Assistants, and Autonomous Agents) that are emerging. Potential applications of generative AI are highlighted through five lenses: Individuals, Functions, Organizations, Industries, and the Ecosystem. Finally, to navigate this journey effectively, the book suggests that organizations should follow a learning and adoption process to develop organization-wide capabilities that progress through predictable stages. Starting with Superusers, the process continues through Diffusion, Experiments, and finally, Scaling up to the whole organization. This approach is aimed at allowing readers to integrate and apply generative AI incrementally, thereby reducing risk and enhancing understanding.

In summary, the key frameworks detailed in the book are:

1. Event History Analysis Framework: Three Strategies for Generative AI
2. Generative AI Product Architectures: Content Generators, Copilot Assistants, and Autonomous Agents
3. Five Application Lenses: Individuals, Functions, Organizations, Industries, Ecosystem
4. A Generative AI Adoption Process: Superusers -> Diffusion -> Experiments -> Scale

These frameworks serve as a roadmap to navigate the ambiguity in the promising generative AI landscape, empowering individuals and organizations to seize the opportunities that this revolutionary technology presents.

Chapter 2

—

Event Horizon Strategy: Managing Uncertainty Around Generative AI

"People call [AI Superintelligence] the Singularity, and that's probably a good way of thinking about it. It's singular—it's hard to predict, like a black hole: what happens past the event horizon?" – Elon Musk[28]

"These [AI] tools will help us be more productive, healthier, smarter, and more entertained." – Sam Altman[29]

"The singularity will either be really successful, in which case we're going to have the biggest boom ever, or it is probably going to blow up" – Peter Thiel[30]

---

[28] Musk, E. (2020). Elon Musk Podcast Transcript, *Joe Rogan Experience.* May 7, 2020

[29] O'Donnell, J. (2024). Sam Altman says helpful agents are poised to become AI's killer function. *MIT Technology Review.* May 1, 2024 https://www.technologyreview.com/2024/05/01/1091979/sam-altman-says-helpful-agents-are-poised-to-become-ais-killer-function/

[30] Schulman, A. (2009). Peter Thiel on the Singularity and economic growth. *The New Atlantis.* October 4, 2009

There are numerous viewpoints on the mechanisms and impacts of generative AI. As a scholar in management, my lens is primarily organizational. Generative artificial intelligence's actual potential, in my opinion, resides in its application by people and teams both inside and outside of companies. After all, the most significant transformations in sectors are sparked by the cooperation of bigger groups of individuals and the resources gathered by these companies. Strategic giants in modern society, organizations have power over people, other companies, and whole industries. These entities are entrepreneurial by nature when they try fresh and creative projects. While many companies are happy to copy past behavior, new ideas can help firms build value. This, in turn, dictates the realm of possibilities for individuals, paving paths for jobs and careers, endorsing specific ecosystem tools, and ushering in dominant industry trends.

The Technology Strategy View: AI S-Curves and Capabilities

Technology Trajectories

One framework stands above the others for academics in management trying to demystify the dynamics of technologies, their progress, and their rippling effects: the idea of a technology trajectory. Grasping this idea mostly depends on realizing that technologies are not unchangeable. Investing, efforts, and

---

https://www.thenewatlantis.com/futurisms/peter-thiel-on-singularity-and-economic

inventions supported by scientists, engineers, and inventors help technologies to be dynamic, changing, and advancing. As such, what could be a known fact about technology now could be totally changed tomorrow? Constant change results from unrelenting creative endeavors and attempts to build on earlier innovations. Organizations and sometimes individual users choosing to employ these technologies have also to look ahead and consider how these tools will evolve.

But precisely what is a technology trajectory? Fundamentally, it is the evolutionary road map of a certain technology across time. Naturally, one only appreciates a technical trajectory in retrospect. Navigating a multitude of conceivable technology courses, deciding what is more likely, and laying bets that reduce technical risk presents a difficult challenge for users and inventors. Users generally gravitate for technology with promise and consistent developments. Conversely, innovators try to avoid sinking their efforts into activities that might finally be abandoned by focusing their energies and resources on specific paths over others. Many of the envisaged paths might fade or stagnate. Users and inventors struggle with a common question: which technologies will fly, how far they will advance, and the degree of their final widespread acceptance?

Path Dependence

The journey of technology evolution is not a free flight of fantasy. Rather, it is anchored in the present and limited by contemporary tools and resources. Burst with ideas, inventors generally start their creative process by building on already-existing solutions—improving, adjusting, and inventing to produce fresh versions. Basically, the foundation of tomorrow's discoveries is today's rock. Scholars examining technological trajectories refer to this as "path dependence." Simply said, choices and actions taken in the past either help or hinder our existing and future technical capacity. This continuity lends a semblance of predictability to a technology's trajectory, especially if one can decode the constraints and understand their ramifications.

Path dependence does not just cast a shadow on the future; it also shapes the present. By its very nature, path dependence can restrict the range of possibilities within a technology trajectory. Thus, inventors may deliberately choose to enable and constrain the future, leading to strategic decisions to reach particular objectives.

A classic example that illustrates the tenacity and implications of path dependence is the tale of the QWERTY keyboard. Now found everywhere, the QWERTY keyboard has an interesting background. Early typewriters sometimes had a problem where typebars would jam when neighboring letters were typed quickly one after the other. Typewriters were purposefully set with keys in the QWERTY configuration to minimize jams by spacing often used letter combinations far apart from one another. Although more effective layouts were later developed and engineers found techniques to minimize typewriter jamming, typewrites became the de facto standard as typists became competent with this layout. The QWERTY layout endured despite the initial design goal—preventing jams— becoming obsolete with new technology—because of the great number of trained typists and the large expenses connected with retraining. This illustrates how initial conditions and decisions, even if they arise from transient challenges, can firmly cement a technology's trajectory for years, or even centuries, to come.[31]

The issue with path dependence is its unpredictability— the apparent trajectories or their constraints can change over time. If always successful, a dominant technology path can eclipse possible new rivals, therefore rendering them less viable. The inverse can sometimes happen; a dominant road may disappear and an array of new, interesting roads emerges. Organizations that have greatly committed in the beginning trajectory may find themselves grounded to a sinking ship when this occurs. Under such conditions, it would have been wiser to take a more cautious attitude: wait and observe.

Still, patience has its negatives as well. Some technical routes, thought to have reached a plateau, can shockingly show incremental improvements that will be relevant for considerably more than expected. One moving example is the development of optical photolithography tools, which are vital for semiconductor chip manufacturing. Over numerous decades, this equipment improved surprising industry estimations in spite of continuous predictions of its approaching stagnation.

This scenario is reminiscent of the "Friedman unit," in which a New York Times columnist declared repeatedly that the following six months would be vital in the Iraq war, a prediction rehashed across several years. Some observers contend artificial intelligence has had its own "Friedman unit." Almost forty years have seen many forecast that human-equivalent artificial general intelligence is always "20 years away." The jury is out on this forecast as we will be discussing later.

---

[31] David, P. A. (1986). Understanding the Economics of QWERTY: The Necessity of History. In W. Parker (Ed.), *Economic History and the Modern Historian* (pp. 30-49). Blackwell.

A typical mistake is confusing the popular acceptance of items that profit from technological breakthroughs with technical progress—improvements in the intrinsic features of a technology trajectory. Usually, the two show a clear lag. User adoption depends on marketing activities, manufacturing capacity, and strategic alliances, and occasionally even requires creative business models to get the technology to the market; it does not only follow technological milestones.

Still, path dependency is not intrinsically bad. As was already mentioned, it allows technological paths some regularity. Developers also usually benefit from building on tested foundations at lower cost. Making the correct technological gamble can pay off handsomely, maybe even resulting in a competitive edge challenging for rivals to copy. Path dependency often helps users, on their side. Often a benefit are predictable improvements along with declining technology access prices. Usually, this predictability follows a typical form as will be discussed below.

The essence is still the same, though: many technical fields are full of possible paths with associated degrees of uncertainty. Selecting the right course of action becomes a vital choice to maximize probable returns and control expenses. This approach can be expensive even if businesses divide their investments among numerous parallel lines to lower risks. Consequently, most people—personal or business—usually rely on a single path and accept the natural uncertainties.

Technology Trajectories: S-Curves in Generative AI

Generative AI will have a unique trajectory that is difficult to predict. Developers and users will take a variety of pathways as they select where to invest and what technologies to utilize. However, determining the consequences of these decisions may take time. As Nobel winner and MIT economist Robert Solow noted in 1987, "You can see the computer age everywhere except in the productivity statistics."[32] Indeed, the economic consequences of digital technology frequently appear after a large latency time.

Generative AI may have more immediate consequences. Stanford Professor Erik Brynjolfsson notes, "It's a great creativity tool. It's great at helping you to do novel things. It's not simply doing the same thing cheaper."[33] The economic potential of generative AI will expand as AI tools like ChatGPT and Midjourney develop over time.

The trajectory that generative AI takes will be critical. Historically, most technological breakthroughs followed an S-curve pattern. This pattern depicts initial incremental gains that lead to tremendous progress and culminate in slow growth. When plotted over time, it resembles the letter "S". This trajectory can be divided into three phases: the "era of ferment," the "takeoff," and the "maturity" phase. Figure 1 depicts a typical S-curve shape. The curve depicts non-linear growth, appearing to linger for a long time (period of ferment) before rising exponentially when hurdles are overcome (takeoff), and finally tapering as it approaches physical or other constraints (maturity).

---

[32] Solow, R. (1987). We'd better watch out. *New York Times Review of Books*.
[33] Brynjolfsson, E. (2023). *Stanford Professor Erik Brynjolfsson on How AI Will Transform Productivity*. https://www.microsoft.com/en-us/worklab/podcast/stanford-professor-erik-brynjolfsson-on-how-ai-will-transform-productivity

Figure 1: S-curve trajectory of technology performance over time

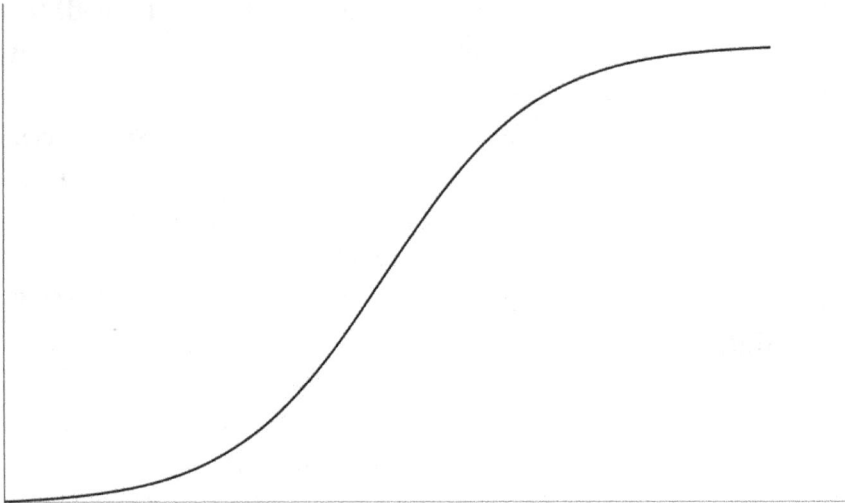

The debate over generative AI's adherence to the S-curve is ongoing. Current data implies a takeoff phase, as seen by both technological prowess (quality of material offered) and user acceptance. Some worry that physical semiconductor limits may hinder progress, especially because generative AI relies heavily on parallel processing GPUs. Furthermore, significant technological advancements in deep learning, transformers, and large datasets may only provide a one-time gain. Others are enthusiastic about future advances in silicon and software. It is unclear if we are approaching the mature phase, although most believe that it is not imminent.

In reality, the long-term evolution of technology-based products may need the development of several S-curves. First, the product may rely on multiple technological components, each with its own set of performance parameters. Personal computers commonly use numerous S-curves for processing power, memory, bandwidth, and energy. Any of these can be a bottleneck for a while before an exponential advance occurs, allowing the entire product performance to improve.

Discontinuities and Multiple S-curves. The second way in which products may involve multiple S-curves is if a given performance dimension is shaped by different technologies that unfold in a sequence. In this way, multiple S-curves may be layered on top of each other. It is well known that breakthrough products often emerge from new S-curves, even as prior generations of products may be disrupted. For example, internet companies like Amazon and Google emerged from a newly emerging web browser S-curve, and Facebook, Instagram, and Tinder were a result of a new S-curve around mobility, as indicated in the graph below.[34]

[34] Zakin, P. (2024). _The biggest consumer internet companies have been started in the wake of technological unlocks._ 2024-08-26 https://x.com/pzakin/status/1827779647725318341

It's 1994 again

August 25 2024

The biggest consumer internet companies have been started in the wake of technological unlocks.

- **Birth of the web browser (Nexus started in 1991, Mosaic in 1993, Netscape in 1994):** Yahoo (1994), Amazon (1994), Ebay (1995), Craigslist (1995), Expedia (1996), Paypal (1998), Google (1998).

- **Expansion of high-speed internet:** LinkedIn (2003), FB (2004), Youtube (2005), Reddit (2005), Twitter (2006), Spotify (2006), Dropbox (2007).

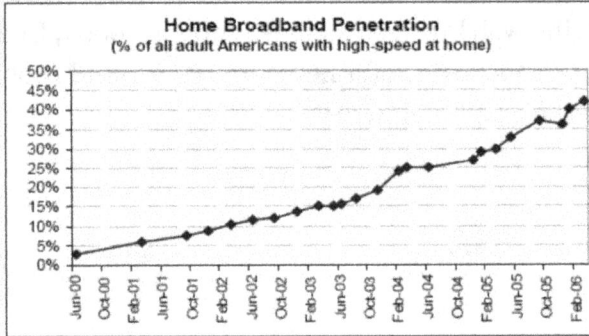

Home Broadband Penetration
(% of all adult Americans with high-speed at home)

- **FB Platform (2007) + iOS App Store (2008):** Airbnb (2008), Pinterest (2009), Whatsapp (2009), Venmo (2009), Uber (2010), Instagram (2010), Snapchat (2011), Tinder (2012).

Indeed, the study of technology strategy is focused on the idea of multiple S-curves, as the transition between S-curves can be difficult for industries and particular firms to make. These are sometimes called "discontinuities"—during moments of discontinuity, organizations face unique challenges of shifting their resources and customer base.[35]   This represents a shift from one S-curve to another.

For example, the popular idea of an "innovator's dilemma" is related to this, as firms may be disincentivized from shifting from an old S-curve with existing customers to a superior new S-curve with new customers.[36] AI arguably has been through four S-curves—expert systems, machine learning, deep learning, and now large language models (LLMs)—each providing a discontinuous increase in performance or capabilities when it emerged, and involving its own S-curve of development.

Locus of Economic Impact of Generative AI. The real economic impact of generative AI might only emerge when businesses harness its potential in actual use cases. Avi Goldfarb, an economist at the University of Toronto, believes the key lies in repurposing this technology to transform businesses, mirroring the computer revolution.[37] Perhaps the most detailed study of use cases is by Anton Korinek, an economist at the University of Virginia and a Brookings Institution fellow, who supports this viewpoint.[38] His study analyzed 25 use cases of generative AI, evaluating its proficiency in tasks ranging from brainstorming and text editing (found to be highly effective) to

---

[35] Tushman, M. L., & Anderson, P. (1986). Technological Discontinuities and Organizational Environments. *Administrative Science Quarterly, 31,* 439-465.

[36] Christensen, C. M. (1997). *The Innovator's Dilemma: When new technologies cause great firms to fail.* Harvard Business School Press.

[37] Agrawal, A., Gans, J., & Goldfarb, A. (2018). *Prediction machines: the simple economics of artificial intelligence.* Harvard Business Review Press.

[38] Korinek, A. (2023a). Generative AI for Economic Research: Use Cases and Implications for Economists. *Journal of Economic Literature, 61*(4), 1281-1317.

coding (decent with support) and mathematical computations (less reliable).

A critical question remains: Will generative AI yield societal dividends, or will its advantages primarily enrich the corporations behind it? MIT economists Daron Acemoglu and Simon Johnson suggest that technological advancements benefit society when nudged by public interest and regulatory interventions.[3] They caution against the unchecked influence of tech magnates, with a particular focus on generative AI, asserting, "Society and its powerful gatekeepers need to stop being mesmerized by tech billionaires and their agenda... One does not need to be an AI expert to have a say about the direction of progress and the future of our society forged by these technologies."

Furthermore, Acemoglu suggests the current trajectory of AI and its creators "are going in the wrong direction." The entire architecture behind the AI "is in the automation mode," he says. "But there is nothing inherent about generative AI or AI in general that should push us in this direction. It's the business models and vision of the people in OpenAI, Microsoft, and the venture capital community." In other words, using generative AI around more creative use cases may be more compelling for society.

Evidence of Technology S-Curve Trajectories Underlying Generative AI

Most technologies follow a predictable development pattern, commonly represented by an S-curve when charting technical performance or customer adoption on the y-axis against time on the x-axis. Progress is modest initially since the technology faces significant hurdles in its early phases. This is the phase during which researchers and developers battle to increase performance, hitting roadblocks that prevent rapid progress. However, once these hurdles are addressed, the technology enters a take-off phase, marked by dramatic performance improvements. This is the steepest part of the S-curve, where the technology's potential is quickly realized, resulting in broad adoption or breakthrough developments. The technology eventually reaches maturity, at which point future advances yield decreasing returns. This phase frequently results in the convergence of the curve, indicating that the technology has attained its full potential or is being replaced by new innovations.

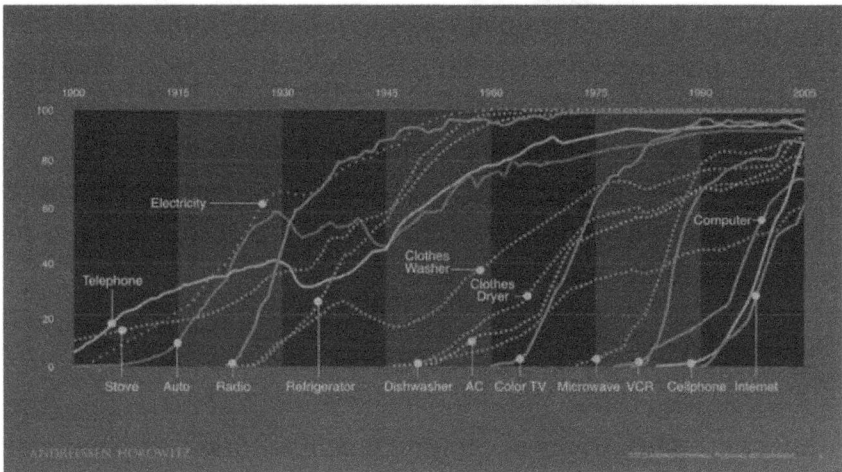

Acceleration of S-Curves with Digital Technologies

As the graph above illustrates, the speed of these S-curves has been accelerating, particularly in the realm of electronics, where some technological breakthroughs have been applied across multiple domains.[39] Digital technologies, in particular, have seen a dramatic acceleration, with adoption curves for innovations like the internet and mobile technologies appearing almost vertical. Generative AI may be the latest manifestation of this trend. The rapid improvement in both the capabilities of generative AI systems and their adoption rates reflect the steep, almost vertical S-curve that is characteristic of many recent digital technologies. This section will detail how these dynamics are unfolding in the context of generative AI.

However, there is a relatively unique aspect to the trajectory of generative AI: the role of very large language models (LLMs). The dramatic improvements in generative AI have been driven by the construction of these extraordinarily large models, which were surprising not only to the public but even to some researchers within the field. It is only by building models of such scale, requiring vast amounts of data and computation, that these exponential capabilities have become possible. These improvements highlight the power of S-curve dynamics, particularly in how change strategic capabilities in an evolving technological landscape.

Scaling Laws

Scaling laws are fundamental to understanding the dramatic performance improvements and exponential capabilities seen in generative AI, particularly as these laws drive the steep acceleration in technology S-curves. Scaling laws describe the relationship between the resources invested into a system—such as computational power, data size, and model complexity—and the resulting performance gains. These laws

---

[39] Chen, A. (2024b). *The mobile S-curve ends, and the AI S-curve begins* 2024-02-23 https://x.com/andrewchen/status/1760698184966504475

show that as we increase the scale of these inputs, we can expect corresponding, and sometimes exponential, improvements in output, thereby driving the rapid ascent seen in the middle of S-curves for advanced technologies. Below, I note major discoveries related to AI scaling laws.

Chinchilla: The Impact of Size. The first major discovery in this area is highlighted by the Chinchilla model, which has had significant implications for our understanding of how to effectively scale AI models. Developed as a relatively small model with only 70 billion parameters, Chinchilla outperformed much larger models by being trained on a far greater number of tokens than had been standard in earlier models. This led to the formulation of the Chinchilla scaling law, which incorporates both parameter count and data size (encoded in tokens) to predict model performance.[40]

The key insight here is that larger models require significantly more data to fully utilize their potential. This is illustrated by a graph below, where Chinchilla, despite its smaller size, outperforms larger models due to the greater volume of training data. This finding underscores the importance of data scaling as a critical factor in the advancement of generative AI, suggesting that outperformance can be achieved not merely by increasing model size but by ensuring sufficient data input.[41]

---

[40] Hendrycks, D. (2023). *Introduction to AI Safety, Ethics, and Society.* Taylor & Francis.
[41] Hoffmann, J., Borgeaud, S., Mensch, A., Buchatskaya, E., Cai, T., Rutherford, E., de Las Casas, D., Hendricks, L. A., Welbl, J., Clark, A., Hennigan, T., Noland, E., Millican, K., van den Driessche, G., Damoc, B., Guy, A., Osindero, S., Simonyan, K., Elsen, E., Rae, J. W., Vinyals, O., & Sifre, L. . (2022). Training Computer-Optimal Large Language Models *Working Paper*. https://arxiv.org/abs/2203.15556

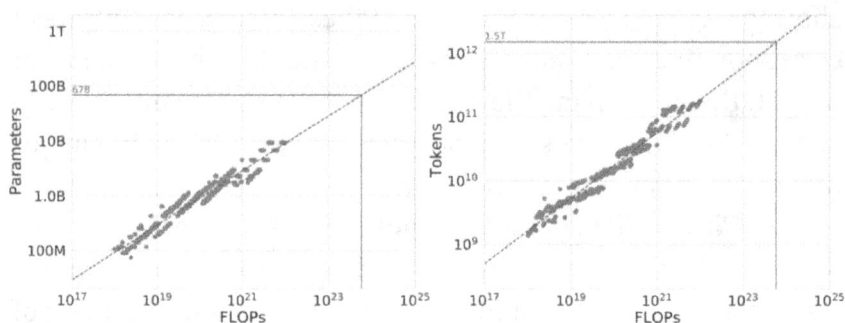

Tradeoffs: Compute, Data Size, and Parameters. With the discovery of generic scaling laws, we see a more comprehensive understanding of the trade-offs between compute, dataset size, and model parameters. These scaling laws show that test loss, a key measure of model error, decreases incrementally as the compute resources, dataset size (measured in tokens), and the number of model parameters increase exponentially. This relationship is visualized in a graph where improvements in these three inputs correspond to a steady reduction in test loss, indicating enhanced model intelligence.[42]

OpenAI's new model series, codenamed "Strawberry" or the "O1" models— including the "O1 Preview" and "O1 Mini," which were designed with chain-of-thought processes to emulate human reasoning—may present a new tradeoff, as a greater share of the computation happens at the point of inference. This may imply a new set of inference-related scaling laws, such that if more tokens and greater computation can be applied to problems, many reasoning problems may be solvable.[43]

---

[42] Kaplan, J., McCandlish, S., Henighan, T., Brown, T. B., Chess, B., Child, R., Gray, S., Radford, A., Wu, J., & Amodei, D. (2020). Scaling Laws for Neural Language Models. *Working Paper*. https://arxiv.org/pdf/2001.08361
[43] OpenAI. (2024). *Learning to Reason with LLMs.* September 12, 2024 https://openai.com/index/learning-to-reason-with-llms/

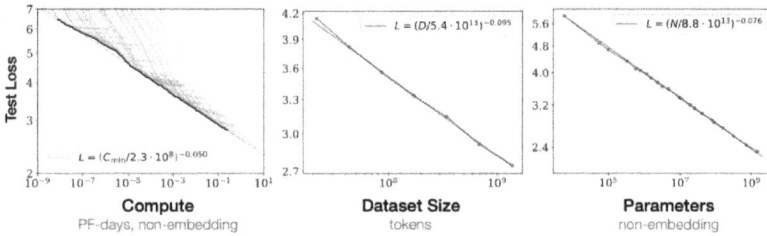

These scaling laws have profound implications: They illustrate that improvements in processing power, data accessibility, and model dimensions all positively influence the capabilities of LLMs. These universal scaling laws offer a framework for methodically improving model performance and have emerged as a fundamental principle in the advancement of progressively complex AI systems.

Scaling Laws in Many Models. These scaling laws have been validated over several model types, encompassing transformers and other architectures. As the quantity of parameters in these models increases exponentially, test loss is significantly reduced, as illustrated in another graph. This validation across several model types strengthens the reliability of scaling laws and indicates that advancements seen in generative AI are not confined to particular designs but may be generalized across different models.[44] The persistent trend of performance improvements via parameter scaling underscores

---

[44] Tom B. Brown, B. M., Nick Ryder, Melanie Subbiah, Jared Kaplan, Prafulla Dhariwal, Arvind Neelakantan, Pranav Shyam, Girish Sastry, Amanda Askell, Sandhini Agarwal, Ariel Herbert-Voss, Gretchen Krueger, Tom Henighan, Rewon Child, Aditya Ramesh, Daniel M. Ziegler, Jeffrey Wu, Clemens Winter, Christopher Hesse, Mark Chen, Eric Sigler, Mateusz Litwin, Scott Gray, Benjamin Chess, Jack Clark, Christopher Berner, Sam McCandlish, Alec Radford, Ilya Sutskever, Dario Amodei. (2020). Language Models are Few-Shot Learners. *Working Paper.* https://arxiv.org/abs/2005.14165

the possibility for ongoing exponential advancements in AI capabilities, contingent upon the continual growth of the foundational inputs—compute, data, and model size.

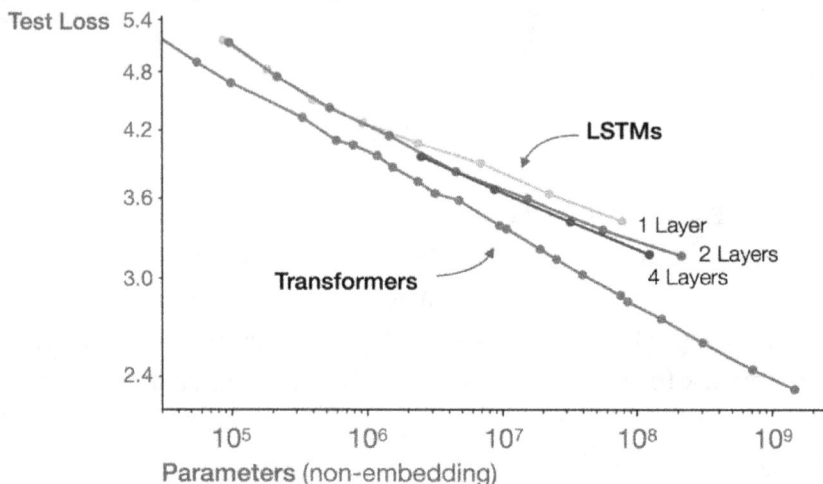

The discovery and validation of these scaling laws lead to a critical conclusion: the S-curves of technological advancement in generative AI, characterized by dramatic exponential improvements, are driven by the continued growth of these underlying inputs. As organizations and researchers push the boundaries of compute, data availability, and model complexity, we can expect to see further steep ascents in performance, continuing the trajectory of rapid innovation in generative AI.

Technological Inputs to LLM S-Curves

The inputs driving the S-curves of LLM performance—computation, context windows, and investment—are accelerating dramatically as companies race to achieve the exponential performance improvements in AI capabilities. This surge in inputs is reflected in several key metrics, each of which contributes to the steep ascent in the intelligence of LLMs.

Compute: FLOPs, Context Windows, and GPUs

Initially, we describe the exponential growth of training computation, as measured by floating-point operations per second (FLOPs), across different models introduced from 2010 to 2024. FLOPs measure the computational requirements for training an AI model, offering a consistent metric for comparing the computational demands of various models. The graph below illustrates the remarkable escalation in FLOPs over time, emphasizing a three-orders-of-magnitude rise in the average model's training computation from 10^{20} FLOPs in 2020 to 10^{24} FLOPs in 2024. For instance, OpenAI's computational investment for training GPT-4 in 2024 was about two orders of magnitude higher than that utilized for training GPT-3 in 2021.[45]

---

[45] AI, E. (2024). *Notable AI Models.* Epoch AI. 2024-07-10
https://epochai.org/data/notable-ai-models

Training compute of notable models

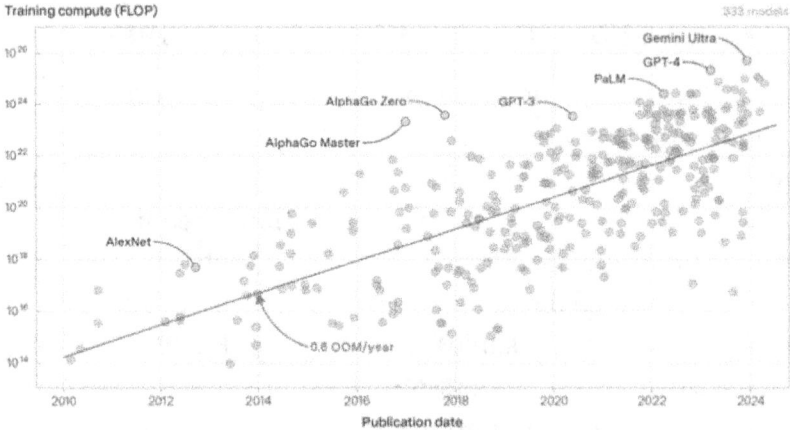

It is crucial to note that compute itself is advancing along its own S-curve, with its technical performance improving dramatically over time. However, in the context of LLMs, compute serves as a critical input, and its rapid increase is a significant determinant of AI performance. The dramatic escalation in FLOPs underscores the commitment of leading companies to push the boundaries of AI capabilities by investing heavily in computational resources.

Another critical input in LLM performance is the size of context windows, which determine how much information an LLM can remember and use in generating responses. A larger context window enables the model to produce more complex, subtle, and accurate inferences, making it a vital factor in enhancing AI performance. The graph depicting context window sizes over time shows remarkable growth, with earlier models like Mistral limited to 8,000 or 32,000 tokens at most; current models, on the other hand, routinely handle hundreds of thousands of tokens. Some of the latest models can even manage context windows with millions of tokens. Retrieving from even trillions of tokens may be possible.[46] This expansion in context window size is expected to continue as compute and memory capacities increase, further enhancing the complexity and utility of LLMs.[47]

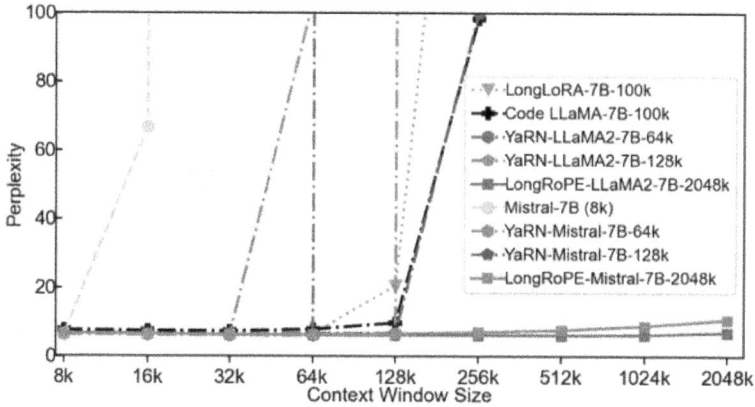

The surge in these capabilities is largely driven by large investments in computational resources by big tech companies in underlying technologies that are themselves improving over time. This investment, itself an input to the S-curves of AI performance, is accelerating at an extraordinary rate. A graph illustrating current and projected investments in AI

[46] Borgeaud, S., Mensch, A., Hoffmann, J., Cai, T., Rutherford, E., Millican, K., van den Driessche, G., Lespiau, J. P., Damoc, B., Clark, A., de Las Casas, D., Guy, A., Menick, J., Ring, R., Hennigan, T., Huang, S., Maggiore, L., Jones, C., Cassirer, A., . . . Sifre, L. (2021). Improving language models by retrieving from trillions of tokens. https://arxiv.org/abs/2112.04426
[47] AK. (2024). *Microsoft presents LongRoPE Extending LLM Context Window Beyond 2 Million Tokens Large context window is a desirable feature in large language models (LLMs).* 2024-02-22 https://twitter.com/_akhaliq/status/1760499638056910955

infrastructure shows that while current spending is in the hundreds of billions, it could easily reach trillions by 2028 if the current trajectory continues. This scale of investment involves ramping up the production of AI chips from millions to hundreds of millions and increasing power consumption from the current 1-2% of U.S. capacity to potentially 20% or more.[48] The scale and speed of this investment reflect the intense competition among companies to achieve leadership in AI capabilities.

Year	Annual investment	AI accelerator shipments (in H100s-equivalent)	Power as % of US electricity production	Chips as % of current leading-edge TSMC wafer production
2024	~$150B	~5-10M	1-2%	5-10%
~2026	~$500B	~10s of millions	5%	~25%
~2028	~$2T	~100M	20%	~100%
~2030	~$8T	~100s of millions	100%	4x current capacity

One final indicator of this trend is the exponential growth in NVIDIA's datacenter revenue, which is closely tied to the demand for outsourced training and inference compute. Many smaller companies, unable to invest in their own AI capital expenditures (CapEx), are turning to outsourcing as a way to access computational resources. NVIDIA is a main beneficiary of this, as a provider of datacenter infrastructure. The graph below shows NVIDIA's datacenter revenue as a dramatic hockey-stick increase, starting in 2022, with revenue

---

[48] Aschenbrenner, L. (2024). Situational Awareness: The Decade Ahead. https://situational-awareness.ai/wp-content/uploads/2024/06/situationalawareness.pdf

soaring to tens of billions of dollars.[49] This trend underscores the rapidly growing demand for high-performance computing infrastructure to support AI development.

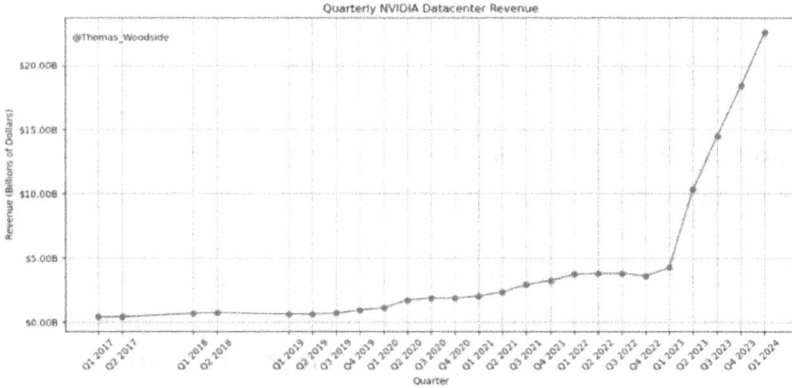

Quarterly NVIDIA Datacenter Revenue

Taken together, these data points illustrate the exponential investments in compute, power, and data potential that underlie the S-curves of LLM intelligence. As companies continue to scale up these inputs, the rapid improvements in AI capabilities are likely to persist, driving further breakthroughs and pushing the boundaries of what is possible with generative AI.

Model Efficiency: Compute, Size, and Cost

A further strong indication of the swift advancement in LLM intelligence is the significant enhancement in model efficiency. This expansion is not only propelled by significant augmentations in computational capacity, energy, and data inputs. Significant algorithm advancements and other forms of

---

[49] Woodside, T. (2023). *Updated graph of NVIDIA datacenter revenue, with the latest quarter.* . 2023-11-25
https://twitter.com/Thomas_Woodside/status/1728195339331751955

LLM research have also empowered these models to accomplish more with fewer resources. This efficiency is essential for maintaining the exponential advancements in AI capabilities, especially when resources like computing power and data become increasingly limited.

Effective Compute. One of the best measures of this efficiency is "effective compute," a metric that considers the computational resources necessary to get a specific level of model performance, incorporating both raw computational power and the efficiency of the utilized algorithms. The graph below illustrates the exponential enhancements in effective computing compared to 2014 benchmarks, with models such as GPT-2 in 2021 and the Chinchilla models in 2022 seeing significant advancements. These enhancements indicate that, in general, computational efficiency is doubling roughly every eight months.

Efficiency doubles roughly every 8 months

Effective compute (relative to 2014)

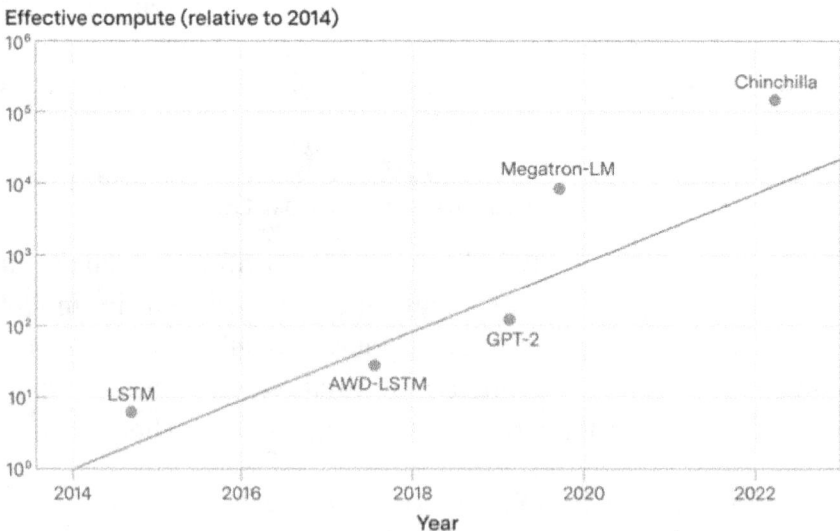

Indeed, efficiency improvements have been achieved in both training and inference, as evidenced by the decreasing cost of inference over time. Some refer to this as "intelligence too cheap to be metered," implying that inference could become ubiquitous. The swift advancement allows intelligence to be applied broadly, and newer models to attain more intelligence with reduced resources.[50]

Relative (inference) cost of ~50% performance on the MATH benchmark

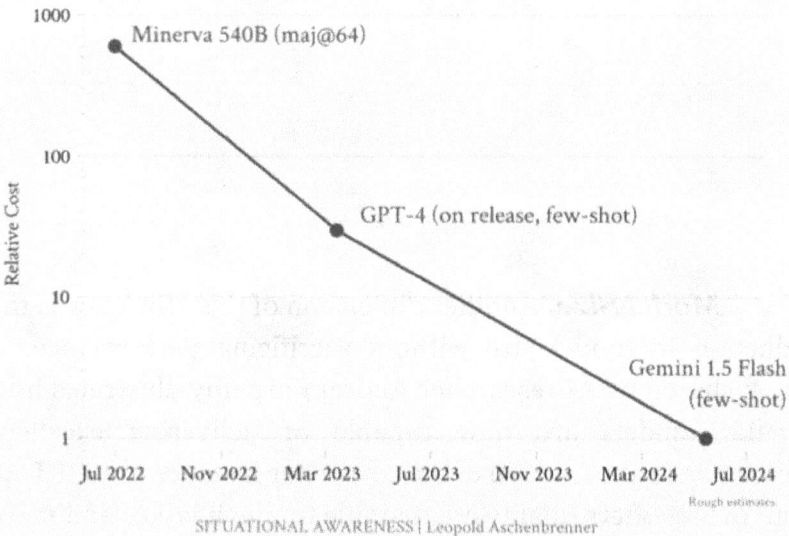

Token Cost. Another indication of cost reduction is in the cost of tokens itself— for example, token cost from GPT-4

---

[50] Aschenbrenner, L. (2024). Situational Awareness: The Decade Ahead. https://situational-awareness.ai/wp-content/uploads/2024/06/situationalawareness.pdf

alone has dropped by multiple order of magnitudes in the last year alone. As Venture Capitalist Elad Gil described, the cost for 2 million tokens decreased from \$180 to \$0.75 in two years, reflecting a 240x reduction in cost.[51]

Token Cost of GPT-4 level models over time

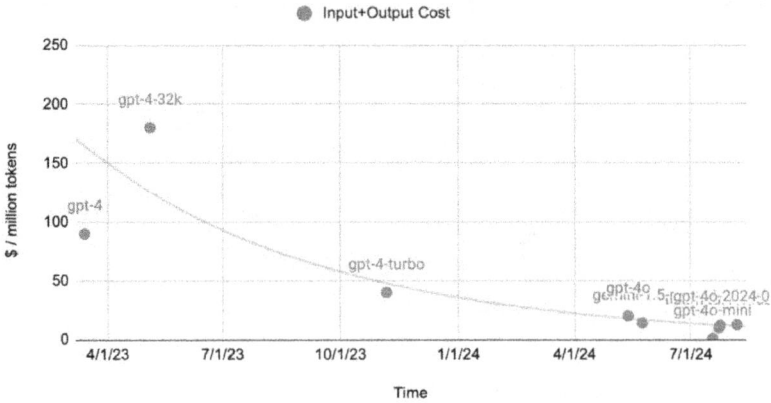

● Input+Output Cost

Cost for 2 million tokens (input+output) decreased from \$180->\$0.75 in 2 years. 240x cheaper

Model Size. Another dimension of this efficiency is the reduction in model size without sacrificing performance. A graph shared by AI researcher Andrej Karpathy illustrates how smaller models are now capable of delivering excellent performance at a fraction of the cost.[52] For instance, the GPT-4o mini model offers quality comparable to much larger and more expensive models like Llama 3. This reduction in size is particularly important as it decreases the computational and memory footprint, enabling these models to be deployed on

---

[51] Elad, G. (2024). *Cost of 1m tokens has dropped.* 2024-08-25
https://x.com/eladgil/status/1827521805755806107
[52] Karpathy, A. (2024). *LLM model size competition is intensifying...
backwards!* July 19, 2024
https://x.com/karpathy/status/1814038096218083497

smaller devices such as smartphones. This capability opens new possibilities for AI applications, making advanced generative models more accessible and versatile.

MMLU vs. Price, Smaller models

MMLU: General reasoning quality benchmark, Price: USD per 1M Tokens

■ Most attractive quadrant

■ GPT-4o Mini ■ GPT-3.5 Turbo ▨ Gemini 1.5 Flash ▨ Llama 3 (70B) ▨ Llama 3 (8B) ▨ NeMo
▨ Mistral 7B ▨ Claude 3 Haiku ▨ Command-R ■ Reka Edge

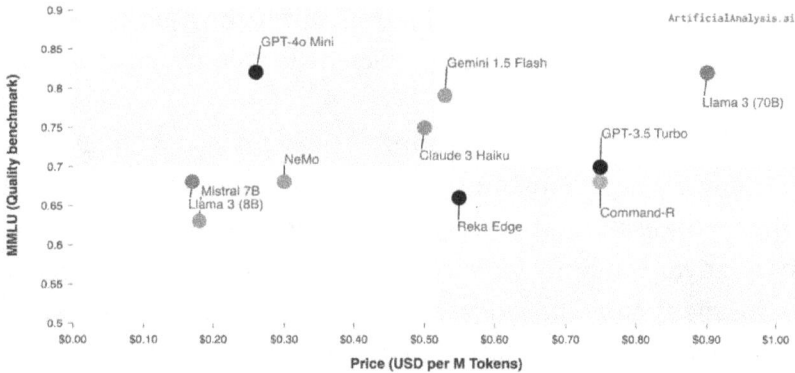

Another Measure: Emergent Capabilities? Some researchers suggest that the true measure of LLM capability is whether unforeseen properties emerge from these systems.[53] These emergent properties may include the capacity to conduct scientific research[54], do mathematical proofs[55], or perhaps do

[53] Wei, J., Tay, Y., Bommasani, R., Raffel, C., Zoph, B., Borgeaud, S., Yogatama, D., Bosma, M., Zhou, D., Metzler, D., Chi, E. H., Hashimoto, T., Vinyals, O., Liang, P., Dean, J., & Fedus, W. (2022). Emergent abilities of large language models. *Working Paper.* https://arxiv.org/abs/2206.07682
[54] Boiko, D. A., MacKnight, R., & Gomes, G. (2023). Emergent autonomous scientific research capabilities of large language models. *Working Paper.* https://arxiv.org/html/2304.05332
[55] Tao, T. (2024). Embracing change and resetting expectations. *AI Anthology.* https://unlocked.microsoft.com/ai-anthology/terence-tao/

deductive reasoning and causal logic itself[56]. The development of deductive reasoning and causal logic has been an area in which generative AI has been critiqued.[57] For example, generative AI famously has difficulties with simple operations like counting the number of times the letter "r" appears in words like *strawberry*. These difficulties in simple logical tasks hold generative AI back from being the most useful tool for strategic decision making, as described later. But it must be noted that this is an area of active computer science research, so this could change quickly.[58]

More broadly speaking, these improvements in efficiency are key to sustaining the exponential growth observed in the S-curves of LLM performance. By getting more out of less, these advancements ensure that the rapid pace of AI development can continue even as the industry faces potential limitations in raw computational power and data availability. This trend towards greater efficiency accelerates the evolution of LLMs and broadens their applicability, making high-performance AI more widely available and cost-effective.

Capabilities: Benchmarking with Activities and Tests

What is most impressive about LLMs is what they can achieve with the intelligence derived from their exponentially increasing resources. These capabilities, akin to the inputs that

---

[56] Huang, J., & Chang, K. C. C. (2023). Towards reasoning in large language models: A survey. https://arxiv.org/abs/2212.10403
[57] Felin, T., & Holweg, M. (2024). Theory is all you need: AI, human cognition, and decision making. https://ssrn.com/abstract=4737265
[58] Huang, J., & Chang, K. C. C. (2023). Towards reasoning in large language models: A survey. https://arxiv.org/abs/2212.10403

influence them, also exhibit S-curve dynamics, characterized by rapid performance enhancement over time, frequently surpassing expectations and benchmarks.

Benchmarking Technical Tasks

One of the clearest indicators S-curve dynamics in LLM capabilities is the significant enhancement in executing particular technical tasks, although at differing rates of progress. The 2024 Stanford University AI Index Report offers a comprehensive study of this advancement. The graph demonstrates that AI models have surpassed human benchmarks in fundamental activities, such as image categorization and visual reasoning, prior to achieving similar results in more intricate tasks, including multitask language comprehension and competitive mathematics. Although AI models exceeded human performance in image-related tasks several years ago, they are just now nearing human-level proficiency in more complex linguistic and mathematics skills, following a recent period of significant advancement

.

MAN VS. MACHINE

AI Models Are Improving Every Year

AI Technical Performance [Selected measures, 100% = human baseline]

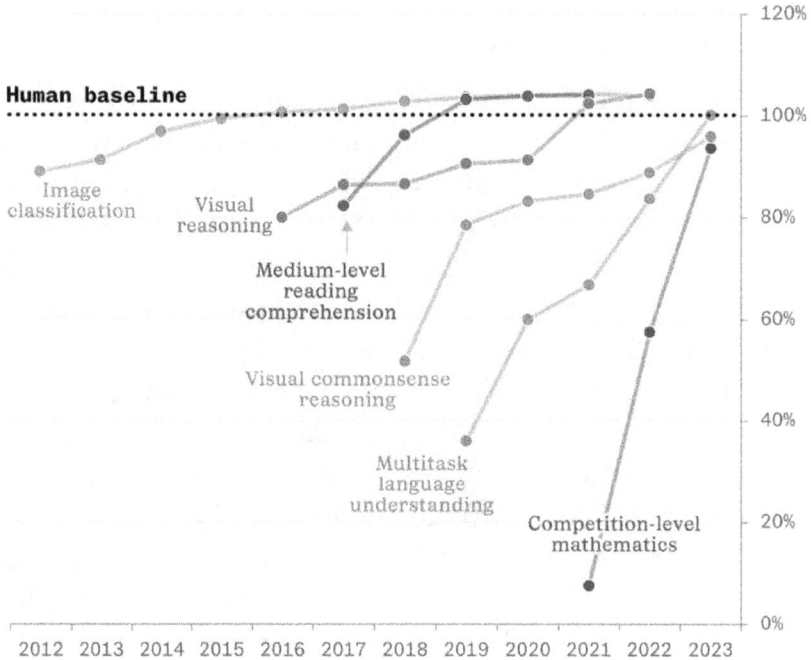

Human baseline

Image classification

Visual reasoning

Medium-level reading comprehension

Visual commonsense reasoning

Multitask language understanding

Competition-level mathematics

120%
100%
80%
60%
40%
20%
0%

2012 2013 2014 2015 2016 2017 2018 2019 2020 2021 2022 2023

CHARTR

Source: Stanford University AI Index Report 2024

Benchmarking Cognition

Another study reinforces this pattern, demonstrating similar outcomes across many disciplines. Tasks associated with perceptual recognition—such as handwriting, voice, and picture recognition—experienced substantial advancements earlier, whereas tasks like reading comprehension and language understanding have just lately made comparable progress. The graph from this study illustrates a distinct trajectory indicating that AI's capacity to process and comprehend language has swiftly progressed, narrowing the disparity with prior advancements in perceptual tasks. This development highlights

46

the adaptability of generative AI systems; even with an 80% efficacy across diverse tasks, their multifaceted talents render them valuable in numerous functions, from fundamental information retrieval to intricate problem-solving.[59]

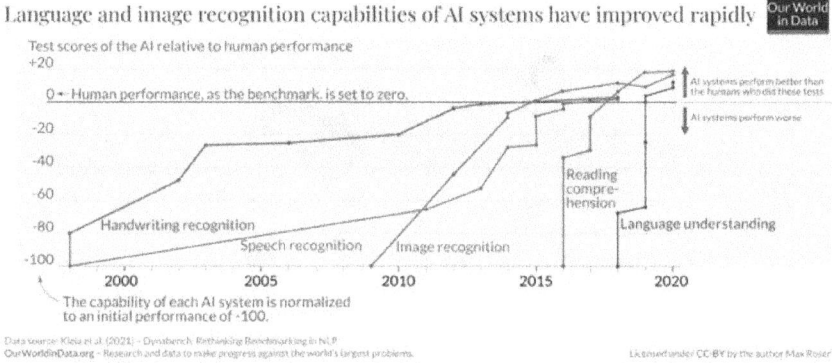

Language and image recognition capabilities of AI systems have improved rapidly

Test scores of the AI relative to human performance

A separate approach has been to give IQ tests to models—for example, one informal test demonstrated that OpenAI's new "O1 Preview" model leveraged chain-of-thought reasoning to achieve an IQ score of approximately 120, which compares favorably to prior models like ChatGPT-4, LLAMA-3.1, and Claude-3 Opus, which typically scored between 60 and 90.[60]

---

[59] Mollick, E. (2023b). *If AI really does plateau at 60-80th percentile of human ability (no sign it will/won't), the impacts may be stabilizing.* . 2023-11-27 https://x.com/emollick/status/1784359024592359587

[60] Lott, M. (2024). *Massive breakthrough in AI intelligence: OpenAI passes IQ 120.* September 14, 2024 https://www.maximumtruth.org/p/massive-breakthrough-in-ai-intelligence

This site quizzes 9 Verbal & 4 Vision AIs every week | Last Updated: 11:08AM EDT on September 14, 2024

IQ Test Results

Reset | Show Offline Test | Show Mensa Norway | ≡

Score reflects average of last 7 tests given

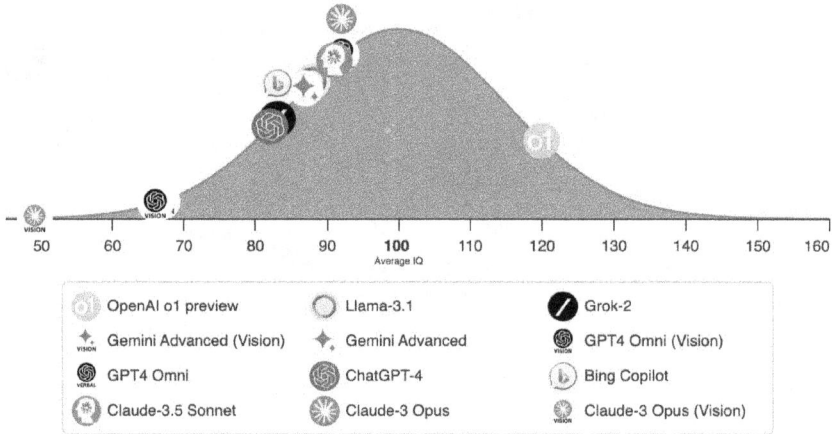

OpenAI o1 preview	Llama-3.1	Grok-2
Gemini Advanced (Vision)	Gemini Advanced	GPT4 Omni (Vision)
GPT4 Omni	ChatGPT-4	Bing Copilot
Claude-3.5 Sonnet	Claude-3 Opus	Claude-3 Opus (Vision)

Benchmarking Real-world Tasks

In addition to these specific technical capabilities, LLMs are also being subjected to rigorous benchmarking tests designed to measure their performance across a range of real-world applications. These include assessments of reading comprehension, general knowledge, code generation, and complex reasoning.[61] Impressively, AI systems have either exceeded or are nearing human benchmarks in all these areas, demonstrating their growing prowess.

---

[61] Douwe, K. (2023). *Plotting Progress in AI.* Contextual AI. 2023-05-23
https://contextual.ai/news/plotting-progress-in-ai/

Test scores of AI systems on various capabilities relative to human performance

Within each domain, the initial performance of the AI is set to - 100. Human performance is used as a baseline, set to zero. When the AI's performance crosses the zero line, it scored more points than humans.

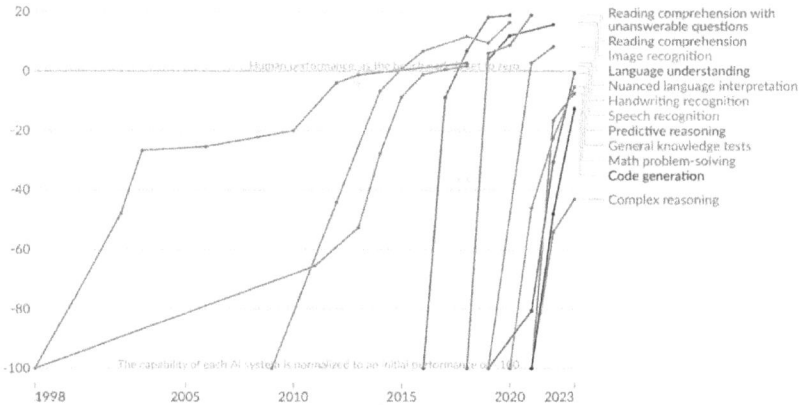

Data source: Kiela et al. (2023)

OurWorldInData.org/artificial-intelligence | CC BY

Note: For each capability, the first year always shows a baseline of ~100, even if better performance was recorded later that year.

Benchmarking Professional Testing

AI models are now excelling at professional exams that are critical for certain occupations, such as the LSAT for law school admissions, the MCAT for medical school, bar exams for aspiring lawyers,[62] and certification exams for other professions.[63]

---

[62] Katz, D. M., Bommarito, M. J., Gao, S., & Arredondo, P. (2023). GPT-4 passes the Bar Exam. https://ssrn.com/abstract=4389233

[63] Aschenbrenner, L. (2024). Situational Awareness: The Decade Ahead. https://situational-awareness.ai/wp-content/uploads/2024/06/situationalawareness.pdf

Performance on common exams
(percentile compared to human test-takers)

	GPT-4 (2023)	GPT-3.5 (2022)
Uniform Bar Exam	90th	10th
LSAT	88th	40th
SAT	97th	87th
GRE (Verbal)	99th	63rd
GRE (Verbal)	80th	25th
US Biology Olympiad	99th	32nd
AP Calculus BC	51st	3rd
AP Chemistry	80th	34th
AP Macroeconomics	92nd	40th
AP Statistics	92nd	51st

SITUATIONAL AWARENESS | Leopold Aschenbrenner

These benchmarks and evaluations demonstrate the potential of LLMs in the domain of capabilities. These models are not merely theoretical constructs with potential; they are real instruments that are currently exhibiting outstanding performance across several tasks. As LLMs advance along their S-curves, they may become increasingly essential in professional, educational, and daily environments, where their capacity to comprehend, generate, and utilize knowledge will be crucial.

Capabilities in the Far Future: AGI and ASI

The most profound future impact of AI lies in the potential arrival of Artificial General Intelligence (AGI), defined as AI systems capable of performing any task better than a human. This milestone will enable AI to be applied to a vast array of problems, constrained only by the availability of computational resources. Predicting when AGI will arrive is challenging, however, and estimates vary greatly. Some experts suggest it could emerge within the next 2 to 10 years.[64] What is most interesting, however, is the trend of estimates—the expected time to AGI has been decreasing, as the ARK invest graph below indicates.[65]

---

[64] Mollick, E. (2023c). *Prediction Markets and AI labs suggest AGI is within planning horizons.* November 28, 2024
https://x.com/emollick/status/1729295140840202481
[65] Management, A. I. (2023). *Ark Investment 2023 Generative AI.* Ark Investment Management LLC. January 31, 2023 Ark Investment Management LLC

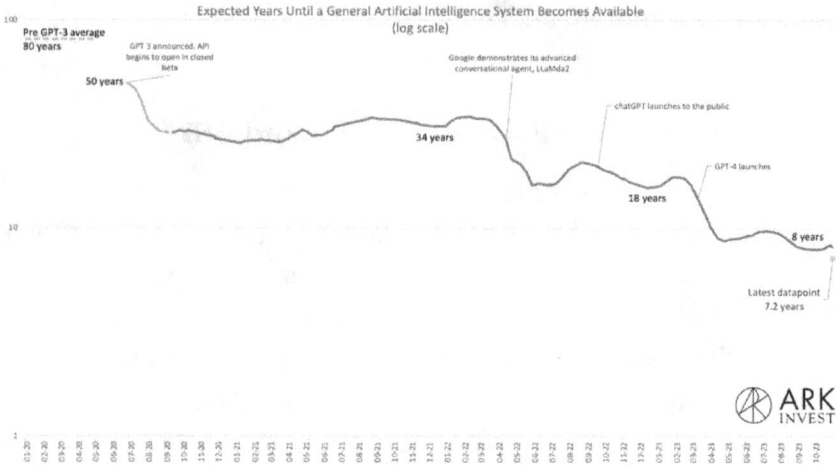

Expected Years Until a General Artificial Intelligence System Becomes Available (log scale)

Source: Metaculus, ARK Invest
Grey lines are derived estimates for time to general purpose AI (strongly formulated) based upon forecasts for a weaker benchmark
For benchmark details see https://www.metaculus.com/questions/5121/date-of-general-ai/, benchmark broadly requires the successful passage of an adversarial two hour Tuning test, broad success on a Q&A knowledge and logic benchmark, and the successful interpretation of and execution complex model car assembly instruction, all within a single system.

Business leaders are generally optimistic about the eventual evolution of generative AI into AGI. According to a recent INSEAD survey, 50% of respondents say AGI will emerge in the next 2 to 10 years, and 8% think AGI could be realized as soon as within the next two years,[66] underscoring business leaders' optimism.[67] These views are echoed by AI scientists, whose estimates for AGI's arrival have steadily shifted closer over time. A significant 10% of AI experts in a 2023 study predicted that AGI would emerge by 2027.[68]

---

[66] Davis, J., & Li, J. B. (2024). Early Adoption of Generative AI by Global Business Leaders: Insights from an INSEAD Alumni Survey *Working Paper*. https://arxiv.org/abs/2404.04543
[67] Davis, J. P. (2024c). What Business Leaders Really Think About Generative AI. https://knowledge.insead.edu/leadership-organisations/what-business-leaders-really-think-about-generative-ai
[68] Katja Grace, H. S., Julia Fabienne Sandkühler, Stephen Thomas, Ben Weinstein-Raun, Jan Brauner. (2024). Thousands of AI Authors on the Future of AI. *Working Paper*. https://arxiv.org/abs/2401.02843

Notable figures in AI have offered perspectives on this timeline too. OpenAI CEO Sam Altman predicted in 2023 that AGI could emerge within the next four to five years, placing its arrival around 2027.[69] Similarly, Avital Balwit, chief of staff to the CEO at Anthropic, stated in May 2024 that she expects AGI to be realized within five years.[70] Leopold Aschenbrenner, another prominent AI researcher, noted in June 2024, "AGI by 2027 is strikingly plausible. GPT-2 to GPT-4 took us from ~preschooler to ~smart high-schooler abilities in 4 years. Tracing trendlines in compute (~0.5 orders of magnitude or OOMs/year), algorithmic efficiencies (~0.5 OOMs/year), and 'unhobbling' gains (from chatbot to agent), we should expect another preschooler-to-high-schooler-sized qualitative jump by 2027."[71] Even Roon, an anonymous yet influential figure in the AI community known to work at OpenAI, expressed in July 2024 that there's a 90% chance AGI will emerge within the next five years.[72] These perspectives highlight the optimistic view prevalent in the industry, suggesting that significant advancements are imminent.

One way to assess the improvement over time is to examine predictions of the technical capabilities of these LLMs in the domain that has been the most studied, software engineering. Of course, a variety of predictions have been made, many of them highly optimistic. For example, one practitioner,

[69] Bajekal, N., & Perrigo, B. (2023). 2023 CEO of the Year: Sam Altman. *Time.* https://time.com/6342827/ceo-of-the-year-2023-sam-altman/
[70] Balwit, A. (2024). My Last Five Years of Work. *Palladium: Governance Futureism.* https://www.palladiummag.com/2024/05/17/my-last-five-years-of-work/
[71] Aschenbrenner, L. (2024). Situational Awareness: The Decade Ahead. https://situational-awareness.ai/wp-content/uploads/2024/06/situationalawareness.pdf
[72] Roon. (2024). *https://x.com/tszzl.* https://x.com/tszzl

Peter Wildeford, notes that although vendors have produced multiple models, the capabilities of OpenAI's ChatGPT product line can be used to compare across time, as indicated in the chart below.[73] In 2023, GPT-4 relied on 2e25 FLOPs, and was roughly capable of being a coding co-pilot. He predicts that in 2026, GPT-6 will have 3e27 FLOPs, and be able to autonomously implement fairly complex programs end-to-end. By 2030, GPT-8 may have 1e29 FLOPs, and perform as a fully automated software engineer, and even autonomously run a small company, an important measure of AGI in the corporate realm.

---

[73] Peter, W. (2024). *What I'm expecting in the next 5-6 years of AI development.* 2024-08-20
https://x.com/peterwildeford/status/1825614599623782490

Model	Release year	Absolute FLOP (median expectation)	Relative FLOPe (median expectation)	Capabilities (very speculative median expectation of capabilities arising from the "Relative FLOPe" amount of FLOPe, which makes this a stronger-than-median but plausible scenario)
GPT-2	2019	4e21 FLOP	5e19 FLOPe	Not much useful
GPT-3	2020	3e23 FLOP	7e21 FLOPe	Copywriting
GPT-4	2023	2e25 FLOP	2e25 FLOPe	Coding co-pilot
GPT-5	2024-2025	5e26 FLOP	1e27 FLOPe	- Automated customer service - more autonomous agents - larger-scale coding assistance (e.g., whole PR)
GPT-6	2026	3e27 FLOP	2e28 FLOPe	- Can design and autonomously implement fairly complex programs end-to-end, though not entire products / companies
GPT-7	2027	1e28 FLOP	2e29 FLOPe	- Fully automated BI analyst - AI might start fully replacing some skilled labor
GPT-8	2030	1e29 FLOP	7e30 FLOPe	- Fully automated software engineer? - Could autonomously run a small company with a few employees? - Potentially coming close to AGI / human-level on all tasks?

After this point, the returns from whether or not – and how – automated development improves AI capabilities itself makes things especially hard to predict, so I cannot hazard much of a guess beyond 2030.

= Estimated from current data, with some uncertainty (e.g., moderately concrete from reporting)
= Forecasted, uncertainty omitted but is considerable (e.g., this is totally made up)

Relative FLOPe → equivalent compute to what X FLOP would get in 2024, adjusting for algorithmic progress

Source: @peterwildeford, personal forecasts and estimations using data from Epoch AI

However, this optimism is not universal. In their book "AI Snake Oil," Arvind Narayanan and Sayash Kapoor believe that much of the excitement around AGI is based on "trend extrapolation," which they call "baseless speculation." This mirrors the many detractors of generative AI who claim it does not understand its outputs.[74] They warn that while models are becoming smaller, they are being trained for more extended

---

[74] Chomsky, N., Roberts, I., & Watumull, J. (2023). Noam Chomsky: The False Promise of ChatGPT.
https://www.nytimes.com/2023/03/08/opinion/noam-chomsky-chatgpt-ai.html

periods of time, potentially leading to diminishing returns and bottlenecks. Furthermore, they point out that meeting professional benchmarks does not always transfer into practical utility, as these studies may have methodological errors, such as contamination of training data.[75]

This more conservative viewpoint is echoed in larger surveys of computer scientists, where estimates for AGI's emergence have historically been more cautious. The 2023 survey, for example, found that many people predict AGI will not be reached until 2047.[76] However, even these predictions are rapidly becoming more accurate, indicating a shift in consensus. Interestingly, prediction markets—a platform where people wager on future outcomes—estimate that AGI will occur by 2033. These markets can provide a more accurate depiction of true consensus since participants are financially motivated to gamble on their genuine ideas about the future.

Yet, even if AGI is accomplished within the next decade, it will only be one stage in AI's evolution. Beyond AGI is the concept of Artificial Superintelligence (ASI), in which AI capabilities greatly exceed those of any human, altogether transcending the limitations of AGI. ASI would usher in a new era in which intelligence, as we know it, is fundamentally redefined. If the transition from AGI to ASI occurs, it will represent the pinnacle of the exponential S-curve dynamics that have fueled AI's fast progress thus far. This transformation is

---

[75] Narayanan, A., & Kapoor, S. (2024). *GPT-4 and Professional Benchmarks.* AI Snake Oil. https://www.aisnakeoil.com/p/gpt-4-and-professional-benchmarks

[76] Katja Grace, H. S., Julia Fabienne Sandkühler, Stephen Thomas, Ben Weinstein-Raun, Jan Brauner. (2024). Thousands of AI Authors on the Future of AI. *Working Paper.* https://arxiv.org/abs/2401.02843

something that the world may look forward to—or prepare for—with cautious anticipation.

General Purpose Technologies and Generative AI

The overarching societal value of generative AI might hinge on the breadth of its applications to challenges faced by individuals and organizations. There is a possibility that LLMs have a narrow application spectrum, perhaps limited to producing marketing content. However, early indicators suggest that LLMs cater to a vast array of use cases, as suggested above. Consequently, some contend that generative AI can be classified as a General Purpose Technology.

Defining General Purpose Technologies

General Purpose Technologies are distinguished by their widespread application across multiple industries. Historically, only a few innovations have achieved General Purpose Technology status, such as electricity, computers, and the internet. These technologies initially cause disruptions, but they eventually result in net positive value due to their numerous beneficial applications across multiple domains.[77]

It is clear that generative AI will transform knowledge tasks that involve language. However, recognizing LLMs as a General Purpose Technology extends beyond linguistic content generation; it is based on the belief that language can effectively encapsulate a wide range of problems, including those unrelated to linguistics.

---

[77] Bresnahan, T. F., & Tratjenberg, M. (1995). General Purpose Technologies: 'Engines of Growth'? *Journal of Econometrics*, 65(1), 83-108.

DALL-E, an image generation LLM, was one notable pivot in the generative AI domain. The idea that visual inference could use language as a "universal interface" was a revelation. These visual-centric LLMs have since undergone progressive improvements.

Historical Precedents and Disruptions

The ripple effects triggered by General Purpose Technologies can be monumental, as electrification showed. Powered predominantly by steam engines, the US, which was the world's leading economy, had manufacturing accounting for half of its GDP by the year 1910. These industries were controlled by conglomerates known as Industrial Trusts. The onset of electrification started with seemingly evident applications, focusing on cost-efficiency. Transitioning from steam to electricity slashed energy expenses for some by a staggering 20% to 60%.

Later, visionaries recognized that reimagining entire factory designs around electrification could amplify these efficiencies, thereby disrupting competitors who did not change their factories. By 1935, electrification had decimated or reduced 40% of the Industrial Trusts that had thrived at the century's start, while the rest declined by average 1/3rd market share. [78] Most economic historians suggest this upheaval was a direct consequence of electrification.[79] Notably, this transformation

---

[78] Goldfarb, B. (2005). Diffusion of general-purpose technologies: understanding patterns in the electrification of US Manufacturing 1880-1930. *Industrial and Corporate Change, 14*(5), 745-773.

[79] Granovetter, M., & McGuire, P. (1998). The Making of an Industry: Electricity in the United States. In M. Callon (Ed.), *The Law of Markets* (pp. 147-173). Blackwell.

spanned a quarter of a century in the US and took even longer in other nations.

If generative AI truly achieves the status of a General Purpose Technology, we might witness similar shifts. Initial advances might focus on labor cost reductions. However, companies that integrate AI first could potentially reap most of the benefits, overshadowing their competitors. It is worth noting that this evolution is likely to be more accelerated than electrification, given that S-curves are hastening, particularly those relying on digital technologies and internet distribution.

Though predicting the future is difficult, it is apparent that a substantial discontinuity is on the horizon, set to impact diverse sectors. The fate of individuals, businesses, and entire industries will hinge on how aligned they are with the evolving trajectory of generative AI.

The Problem of Strategic Decision-making about Generative AI

Research-based frameworks for technology trajectories and General Purpose Technologies are essential tools for making strategic decisions about generative AI. It is critical to understand right away that individuals and businesses are constantly making decisions that are essentially bets on specific technological trajectories. These decisions, whether explicit investments or implicit decisions not to invest, will influence the benefits they derive from generative AI and shape their future strategic options. Describing these trajectories is more than just a chronological account of technological evolution; it allows managers to investigate various strategic paths, even if the true direction of technology is only apparent in retrospect.

However, while these macro-level frameworks provide a satisfying intellectual structure, they frequently fail to provide clear directives for individuals and organizations dealing with specific technologies and products. Context is key. Managers can develop risk-reward strategies within the specific contexts of individual technologies and organizational settings. Given a particular context, there are numerous potential technological trajectories, and it is the manager's responsibility, informed by that context, to determine which trajectories are most likely and beneficial.

The classification of certain generative AI technologies as General Purpose Technologies is advantageous for managers. Managers can draw analogies, estimate technology trajectories, and apply these insights to their specific contexts by studying the early applications of these technologies in various sectors and organizations. Managers can use analogical reasoning across functions, organizations, and industry verticals to identify high-value-added applications. A similar process took place with the spread of electrification and the internet.

However, the generative AI landscape is unique in that it is rapidly expanding and has the potential for enormous but ambiguous impact. As previously stated, even conservative projections indicate significant positive outcomes regarding adoption rates, productivity gains, and revenue generation. In contrast, one can predict generative AI's massive disruptive impact, including job losses, organizational obsolescence, and broader societal challenges.

A common thread running through these scenarios is the profound uncertainty surrounding these predictions. Even though the overall magnitude of generative AI's impact is clear, the specifics remain unknown. The uncertainty surrounding timing and areas of application complicates strategy formulation. Although technology trajectories and cross-case analogies help navigate this fog, they do not precisely direct decisions.

Historical changes caused by General Purpose Technologies provide some insight. Many people underestimated the transformative power of these technological shifts as they occurred. This frequently resulted in a pattern of underinvestment despite occasional instances of excessive enthusiasm that did not pan out. It's worth noting that even these mistakes can be advantageous. Organizations may gain knowledge and insights from these ventures, providing a competitive advantage when the right technological path is eventually identified. Furthermore, certain technological assets or components may lose value or find new applications as the landscape shifts dramatically. Strategists are thus navigating a storm of potentially transformative but uncertain technological trajectories in generative AI. How should it be done?

Event Horizon Strategizing: Managing the Impact and Uncertainty of Generative AI

What is an effective strategy for creating and capturing value from generative AI? I present the "Event Horizon strategy" in this book to grapple with the challenges and opportunities of generative AI. Drawing inspiration from the mysteries of black holes in cosmology, the Event Horizon strategy is defined as a purposeful path of experimentation that ventures into the technology trajectories surrounding these impactful but uncertain technological shifts.

In the world of physics, black holes are viewed as "singularities": regions of infinite density and gravitational pull from which nothing, not even light, can escape. As objects approach a black hole, they reach a boundary called the event horizon. Beyond this point, escaping the black hole's overwhelming gravitational influence becomes nearly impossible. This creates a structure of maximal uncertainty, as what lies beyond the event horizon remains a mystery.

Drawing a parallel, the event horizon of a technological transition is where significant technological possibilities arise. These possibilities, marked by their potentially transformative impact and the profound uncertainty surrounding them, make strategic planning a difficult task. As one moves closer to this horizon, the strategy should focus on gleaning as much information as possible to better position the organization before the "technological singularity" takes place.

Generative AI's similarity to a cosmological singularity is unmistakable. Predictions of AI's impact oscillate between monumental improvements in productivity, ROI, and human well-being, to dire outcomes affecting employment, value distribution, and even the essence of human existence. Despite these projections, we are swimming in a sea of uncertainty. There is unpredictability about these dramatic impacts—AI might overdeliver, underwhelm, or even astonish beyond our wildest estimates. Moreover, the diffusion and applications of AI as a General Purpose Technology remain nebulous. This includes its potential effects across professions or industries and the business models or enterprises that will benefit.

Generative AI may not embody the "infinite" intensity of a black hole, but its ramifications could be profound and even unsettling. There is a growing sense that we are nearing an event horizon in the AI space, where uncertainty is amplifying, yet there is still a window to glean insights about the impending future—at least until a true singularity event, such as the advent of AGI or ASI.

AGI, or "Artificial General Intelligence" is usually defined as a level of intelligence in which AGI could accomplish any tasks a human or animal can perform, whereas ASI or "Artificial Super Intelligence" is a level of intelligence far exceeding human capabilities in nearly all domains. These create events in which human capabilities are rendered irrelevant, or when AI's actions may even be uncomprehensible to human beings. These profound prospects will be explored in subsequent chapters. For now, it is evident that while we might not have created AGI or ASI yet, AI capabilities are skyrocketing, even if the post-singularity landscape is unclear.

But strategists cannot afford to wait. They need to act, make choices, and embark on potential technology trajectories today. Strategies that foster learning and innovation will position companies at an advantage, regardless of whether the singularity is imminent or lies decades ahead.

In its essence, the Event Horizon strategy is a purposeful approach to navigating the generative AI transition. Given the potent outcomes and the enveloping uncertainty, this strategy emphasizes deliberate experimentation. It seeks to traverse technology trajectories in ways that augment both AI-related insights and the enterprise's positioning in terms of products and services. This approach enables organizations to venture near the event horizon, capitalizing on opportunities as the singularity looms closer.

Organizational Inertia as a Strategy

Typically, organizations might adopt one of two polar strategies regarding the generative AI shift. Some might downplay AI's significance, making minimal forays beyond integrating a handful of AI-infused digital tools. This approach, steeped in inertia, has been a recurrent theme in many organizations. In our singularity analogy, these organizations are like those heading straight into the black hole, bypassing the event horizon's explorative potential. Conversely, others might embark on a scattergun approach to learning—dabbling in random AI projects without a cohesive strategy. Such endeavors, though numerous, might not synergize, reducing the likelihood of any significant success. This is comparable to erratic maneuvers around a tiny portion of the event horizon, with the black hole's gravitational might inexorably drawing the organization into its depths.

Purposeful Experimentation as a Strategy

The Event Horizon Strategy I describe in this book underscores purposeful experimentation. It is about charting technology trajectories that genuinely contribute to AI-driven value creation. Envisioning this in the context of a black hole, this strategy amplifies the exploration of the event horizon with elongated, informed trajectories. Pivots and directional shifts are driven by insights gleaned from this journey, stressing both learning and progression. While the singularity might be an inevitable culmination, this strategy seeks to maximize the bounties of generative AI before that defining moment.

Every event horizon strategy shares a few characteristics. First, the strategy should be directed towards a clear direction. While the actual path may diverge based on the learning terrain, the ultimate target remains consistent. Additionally, the path taken should be robust to multiple trajectories. This robustness allows for learning and capability building from various dimensions. This multipronged approach facilitates better adaptation to unforeseen contingencies. It also encourages steady progress through a series of wins, creating momentum. This momentum might be reflected in the achievement of technical or commercial milestones that enhance resource access and inform better decision-making. In essence, the Event Horizon approach to managing uncertainty strives to maximize the exploration of possibilities while exploiting current gains over time.

This strategy is distinct from other approaches to uncertainty management outside the technological singularity context. Entrepreneurship and innovation literature might advise making big bets, creating a diversified portfolio of projects, or continually pivoting between projects. However, these tactics do not align with an event horizon strategy. Making singular big bets might be suitable for predictable, high asset-intensive industries like construction, but can be disastrous amidst high uncertainty. While creating a diversified portfolio may be effective for venture capital investments in startups, it might dilute resources and duplicate efforts. Constant pivoting can also disrupt synergies or inhibit sustained progress. Collectively, these approaches fail to explore the breadth of the technology space adequately and often do not progress towards a definitive objective.

Tesla Motors in Full-self-driving, Robotaxis, and Humanoid Robots

A commendable early example of an event horizon strategy is Tesla Motors and its investment in "real-world AI." This refers to AI capabilities used in the physical realm as opposed to the online domain. Though primarily known as a car manufacturer, Tesla has, for over a decade, honed world-class AI capabilities. Tesla, led by CEO Elon Musk, has set a clear AI technology trajectory. The company planned distinct phases, each aimed at improving AI's tangible role in the real world and marked by commercial goals, which cumulatively led to broader impacts.[80]

Tesla's AI strategy consisted of three main phases. The initial focus in phase 1 was on developing a mass-market electric vehicle with limited autopilot and self-parking capabilities. Crucially, these vehicles had cameras that collected a massive amount of visual data, providing invaluable insights into real-world conditions and serving as a testbed for limited autonomy experiments. The second, subsequent phase revolved around developing Full-Self-Driving (FSD) capabilities. The goal of this current phase would potentially render drivers redundant, paving the way for a robotaxi product line. The ultimate vision, though, culminates in a third phase to develop "Optimus"—a humanoid robot product. This robot would leverage visual and movement systems developed in the FSD phase to enable the robot to undertake any physical human task. Its potential market is estimated to be at least tenfold that of FSD vehicles; if successful, most of Tesla's value would come from real-world AI in humanoid robots. What distinguishes Tesla's strategy is the complementarity across phases. Outcomes from earlier phases benefit subsequent ones, sometimes in unforeseen ways. For instance, the vast dataset from Tesla's EV fleet developed in phase 1 was instrumental for the FSD phase. Musk has described how the neural network did not work well until it had been trained on at least a million video clips. This gave Tesla a big early advantage over other car and AI companies as it had a fleet of almost 2 million Teslas around the world collecting video clips every day.

---

[80] Davis, J., & Yang, D. (2024). Tesla's Real-World AI: Full-Self-Driving, Robotaxis, and Humanoid Robots. *INSEAD Publishing, 10/2024-6903.*

Initial FSD attempts leveraged neural networks for specific tasks, such as lane-following and obstacle avoidance, which were ultimately integrated into driving. These initial systems were competent but not entirely autonomous. However, with advancements in neural networks, Tesla recently made a radical decision to rely solely on them, eliminating vast chunks of hard-coded areas. Musk calls this the "Nothing but nets" approach. The shift was towards learning from billions of clips to replicate human driver responses. The FSD phase is illustrative of the need for adaptability within technology trajectories. Initially, like other AI entities, Tesla depended on NVIDIA's GPUs to train its neural network models. However, these were not optimized for FSD as they are designed to train a large variety of AI models, including LLMs. Tesla's solution was to craft its own DOJO supercomputer cluster with custom silicon, specifically tailored for FSD. This unexpected venture led Tesla to construct one of the world's most potent computing clusters with over 100 Exa-Flops of compute coming online by 2024.

This Tesla case draws attention to some interesting features of an event horizon strategy. The uncertainty of the singularity suggests that during first projects maximum learning is necessary. Still, the decisions should remain flexible and always advance towards general goals. Though sometimes ridiculed, strategic planning is absolutely vital. Though plans are flexible, having a current plan and some alternatives is essential to negotiate technology paths. The approach calls for a whole view that includes possible risks as well as immediate goals.

Strategic Trajectories around AI Singularities: Three Ideal Types

Exploring strategic approaches within the context of an event horizon can be illuminating. A key idea is that the event horizon demarcates the space of high uncertainty after which events and outcomes in a singularity can not be predicated. Assuming the proximity of the event horizon and the inevitability of the singularity's gravitational influence, we can outline three idealized strategic trajectories for leveraging the event horizon.

Direct Descent Strategy

The first strategy entails a direct descent towards the singularity. Given the intense gravitational pull of the black hole, a swift descent is likely to result in destruction for most, with only a few surviving the journey. This involves either a swift, aggressive engagement with generative AI, and/or engaging in activities that are susceptible to replacement with generative AI. For example, certain industry sectors that are now known to rely heavily on tasks readily replaceable by generative AI might find themselves unable to adapt or benefit from generative AI before it fundamentally disrupts their industry. Sectors such as graphic design and copywriting, which depend almost exclusively on content generation, may fall into this category. We might expect full-scale disruption of these industries.

Yet the more interesting case is of those companies that fully embrace generative AI, engaging in activities that aim to accelerate AGI, with few other activities or priorities. Companies like OpenAI and Anthropic, which focus on foundational LLM model creation, are pursuing this strategy—by rushing into generative AI development, they are accelerating the industry towards any AI eventually. Companies like Microsoft and Meta are developing foundational LLM models, but seek to monetize primarily through generative AI-based improvements to existing product-platforms like Office and Instagram. By contrast, companies like OpenAI and Anthropic have so far focused their business models primarily on charging consumers and enterprises for access to foundational models themselves. While it remains to be seen, companies pursuing these models may be engaging in a race to the bottom with ever higher capital expenditures and price wars that result in little profits and only one or a few winners at scale. They may only escape this fate if they develop ancillary products and services that take advantage of their foundational models to provide sustainable value for customers. Both types of companies—those traditional established companies unknowingly transitioning the event horizon and those rushing in head first with advanced model development—may accelerate their decent into an AI singularity in which markets are unpredictable, advantages are unsustainable, and profits are diminishing.

Figure 2a depicts the direct descent technology trajectory, which might be described as accelerating towards the singularity past the AI event horizon, beyond which it is difficult to predict outcomes, profits may be lost, and disruption may be imminent for most firms.

Figure 2a: Direct Descent Strategy in the AI Event Horizon

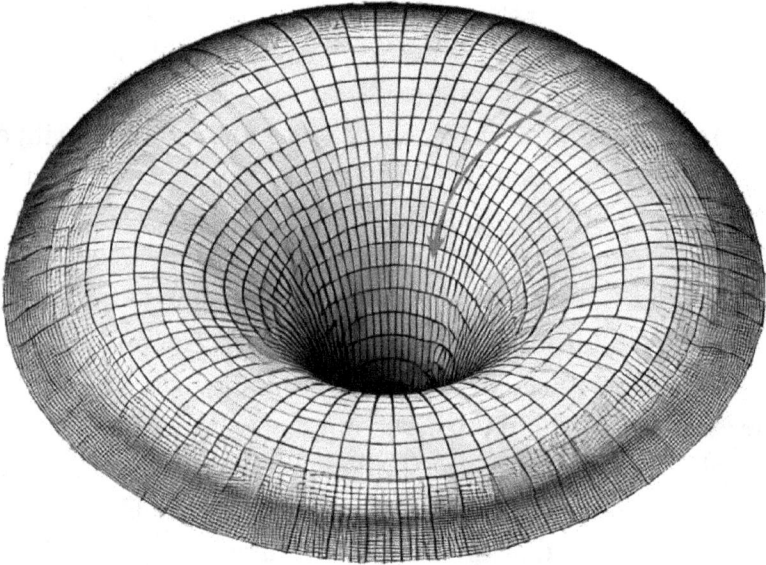

Circumnavigation and Delay Strategy

The second strategic trajectory opts for circumnavigating the outer rim of the event horizon to delay disruption. These entities aim to avoid the singularity for as long as possible, foregoing the benefits of its high energy and acceleration potential, only to be eventually overtaken by the black hole. They may deliberately delay AI adoption, maximizing current business models, and avoiding immediate disruption, often by engaging in minor or incremental generative AI products in a defensive guise. Such companies may get lucky with one or a few of their many pilots of proof-of-concepts, but most of these may be insufficiently ambitious or resisted by organizational inertia. This strategy suits firms in traditional industries not immediately disrupted by generative AI, choosing instead to maximize profits from their existing models for as long as possible. Industries like construction or mining might find this approach suitable, benefiting less from AI in the short term but also avoiding disruption until AI developments related to robotics become prevalent. Such firms many even engage in a few minor AI projects that give some short-term advantages over rivals. However, these firms are unlikely to develop substantial organization-wide capabilities or business units dedicated to AI, preferring to keep AI efforts small. Figure 2b depicts the circumnavigation and delay technology trajectory.

Figure 2b: Circumnavigation and Delay Strategy in the AI Event Horizon

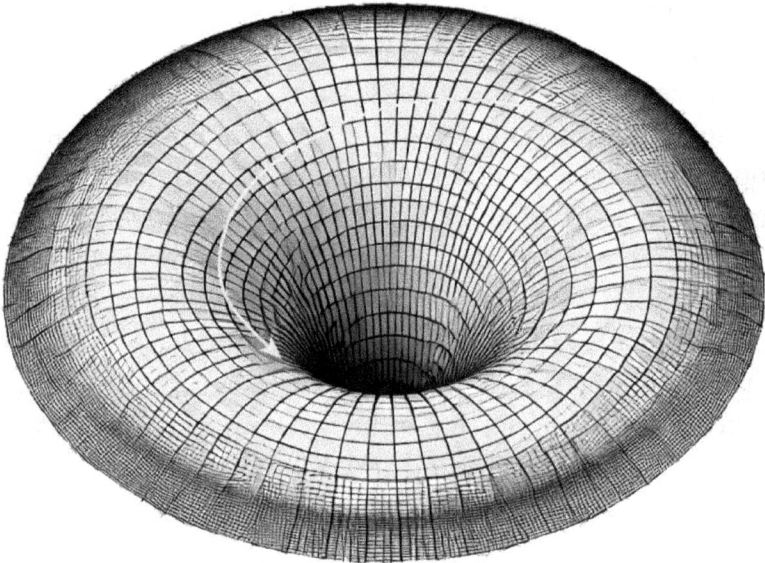

However, it should be noted that in a world of organizations engaging in circumnavigation and delay strategies, the competitive advantages and profits from these activities may actually accrue to consulting service firms like Accenture, McKinsey, or specialized AI platform providers, that enable established organizations to deal with the AI digital transformation. These AI service providers capture the value from enabling established organizations with AI solutions.

A good example is Palantir.[81] Palantir Technologies focuses on enabling established companies to leverage AI effectively, by using the client's proprietary company data to drive value creation. Palantir's Foundry platform plays an underlying role in data management and integration, allowing organizations to analyze and optimize their operations with real-time insights. Palantir recently introduced the Artificial Intelligence Platform (AIP), which focuses on secure use of LLMs within an enterprise security policy. By offering a comprehensive suite that includes both Foundry for data orchestration and AIP for AI deployment, Palantir enables companies to focus on value-added products, services, and operational improvements. This approach enables companies to harness generative AI in a way that aligns with their specific needs, ensuring both innovation and compliance with stringent security standards.

Zig-zagging Sequences Strategy

The third strategic approach seeks a balance between survivability and exploring the event horizon's high energy potential. This strategy involves a zig-zagging trajectory around the event horizon, gaining momentum from areas with strong gravitational pull while maintaining the capability to swiftly recover from unintentional disruption. This strategy involves strategic, phased investment in AI with a focus on learning, with short-term gains building long-term survivability. The key is to link movements together in a sequence where broad capabilities and advantages are built over time. Hence, I refer to this approach as the main "event horizon strategy" throughout this

---

[81] Davis, J. P. (2024b). Palantir Technologies: Enabling AI and Data Science Transformation for Organizations. *INSEAD Publishing, 10/2024-6902.*

book because it aims to maximize benefits in the liminal space between the singularity and the broader universe. Though the singularity might be an inevitable end for these entities as well, they endeavor to explore and extract value from the event horizon to strengthen their position and enhance survivability in the face of the singularity. Companies adopting this strategy may invest heavily in generative AI technologies, ensuring these investments yield sustainable advantages and steering clear of areas where generative AI simplifies task replacement. They orchestrate a series of strategic moves that leverage accelerating generative AI technologies while maximizing survivability ahead of the ultimate singularity of an AGI capable of performing all tasks. Figure 2c depicts the zig-zagging sequences technology trajectory.

Figure 2c: Zig-zagging Sequences Strategy in the AI Event Horizon

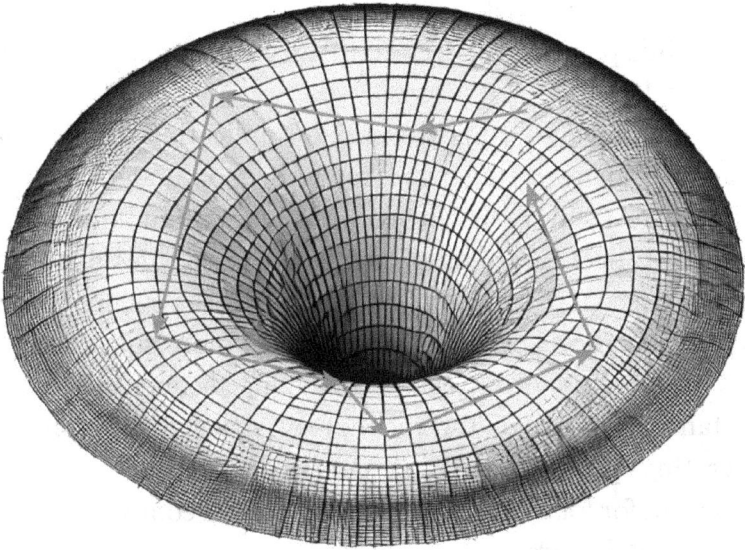

Tesla exemplifies this third "event horizon strategy" of zig-zagging through a sequence of strategic AI investments, each with its own business case but intricately connected to create value across multiple AI domains. The company initially focused on leveraging its EV fleet to gather vast amounts of real-world data, which became the foundation for developing FSD capabilities. As Tesla moved from data collection to implementing FSD, it advanced its neural networks, ultimately relying entirely on machine learning rather than hard-coded algorithms. This progression not only enhanced FSD but also paved the way for future initiatives, such as the robotaxi service and the development of the "Optimus" humanoid robot. Each step in this strategy builds on the previous one, using the insights and technology developed to explore new opportunities while ensuring that Tesla remains adaptable and forward-looking. This approach allows Tesla to extract value from each phase while positioning itself to thrive in the broader context of real-world AI.

Perhaps the quintessential implication of the event horizon strategy is the need for a deeper understanding of generative AI's ever-evolving context. Figure 2d superimposes the three trajectories. Comparing strategies suggests that while direct descent accelerates across the event horizon, circumnavigation and delay is somewhat more sustainable and long lasting. By contrast, zig-zagging sequences explores substantially more of the technology space of generative AI, generating profits based on deeper learning and avoiding disruption for longer periods of time. Table 1 contrasts three key elements of the strategy.

Figure 2d: Three Technology Trajectories in the AI Event Horizon

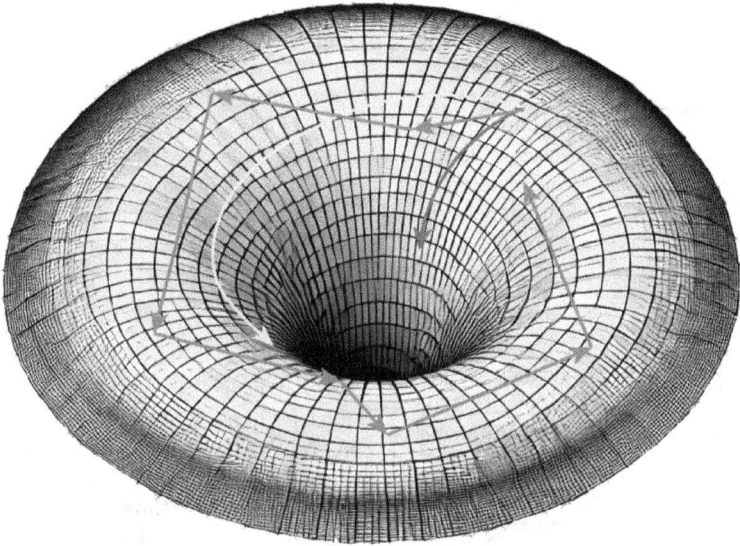

Table 1: Three Event Horizon Strategies

Dimension	Direct Descent	Circumnavigation & Delay	Zig-Zagging Sequences
Core Approach	Swift, aggressive engagement with generative AI, or engaging in activities that are susceptible to replacement with generative AI.	Deliberately delaying AI adoption, maximizing current business models, and avoiding immediate disruption.	Strategic, phased investment in AI, where short-term wins build to long-term survivability. AI learning and adaptability as core decision-factors in each new initiative.
Industries/ Examples	Companies engaged in graphic design, copywriting, or marketing where LLMs are good replacements. Companies like OpenAI and Anthropic focused almost exclusively on foundational model creation.	Traditional industries like construction, mining, or other organizations with competitive advantages not based in knowledge-based resources.	Companies like Tesla that invest in interconnected AI technologies to extract value while preparing for future advancements.
Risks	High risk of full-scale disruption or unsustainable business models; potential race	Risk of being overtaken by AI advancements, loss of competitiveness as AI becomes more integrated	Risk of misaligned investments; requires careful navigation of AI developments to avoid becoming obsolete while

	to the bottom with diminishing returns.	into industry practices. Risk that consulting firms like Accenture, McKinsey, and AI platforms like Palantir capture much of the value from their AI efforts.	maximizing opportu nities.
Benefits	Potential for quick acquisition or early mover advantage, esp ecially in setting industry standards or creating dominant designs in AI.	Stability and extended lifespan of current business models ; time to adapt and prepare for future AI integration.	Maximizes both short-term value and long- term adaptability; sustainable advantages through strategic AI use.
Strategic Flexibility	Low. highly committed to a single trajectory with limited room for adaptation.	Moderate; allows for adjustments over time, but primarily focuse s on delaying significant AI adoption.	High; flexible approach with multiple strategic pivots, allowing for ongoing learning and adaptation as AI evolves.

One of the most important implications of these event horizon strategies is that continual learning about AI in a sequence of experimental projects is the key to taking advantage of generative AI. But simply stating this fact is insufficient—the key to achieving success near the event horizon of generative AI is to resolve uncertainty in a world of continually evolving technologies and unclear value delivery. Finding high-value use cases that individuals, teams, and organizations can utilize to create value with AI is key. Even more than in other areas of strategic management, the devil will be in the details, particularly how generative AI creates enables specific use cases that improve productivity in an organization's workflows. As a result, this book, by necessity, must dig deeper than other typical managerial frameworks and present specific facts and understandings about generative AI, even while these things are constantly evolving. Despite these facts possibly becoming obsolete shortly, it is paramount for strategists to swiftly learn and adapt to the transformative, uncertain singularity that generative AI embodies. There is no substitute for building an intuition from the details and examples that are currently emerging.

Several pivotal contributions are emerging that equip strategists to comprehend the deluge of information about generative AI. This book emphasizes use-cases across various dimensions, from individual benefits to organizational advantages across diverse industry verticals. It proposes canonical event horizon paths applicable across these spectra. Finally, based on these valuable applications, it develops some canonical event horizon strategic paths that work in many individual, functional, and industry contexts.

Chapter 3

—

Generative AI: Affordances and Impact of LLMs

"One doesn't bet against deep learning. Somehow, every time you run into an obstacle, within six months or a year researchers find a way around it." – Ilya Sutskever[82]

The progression of artificial intelligence (AI) has been steady and significant, yet the world has witnessed a transformative leap in its capabilities in recent times. What changed? Historically, AI models, including early neural networks, were predominantly designed and performed well for simple prediction tasks. These models were good at identifying patterns but did not truly *generate* sophisticated novel outputs. The tide shifted when Language Model Large (LLMs) models, notably models like ChatGPT, made their debut, ushering in the era of generative AI.

Generative AI, as its name suggests, is an AI system that can generative extensive and occasionally surprising outcomes

---

[82] Heaven, W. D. (2023). Rogue superintelligence and merging with machines: Inside the mind of OpenAI's chief scientist. *MIT Technology Review*. October 26, 2023

from simple input prompts. While it can produce various outputs—from music compositions to graphic designs—the revolutionary aspect lies in its proficiency with language. Many of these models treat language as a universal representation based on very large databases, which is why they are called Large Language Models or LLMs. In other words, they can, astoundingly, generate images from textual descriptions using systems like DALL-E, Stable Diffusion, and Flux. However, this book's main emphasis will be on their text-generating prowess, where they transform textual prompts into more elaborate, coherent, and contextually relevant text.

Let's delve deeper into LLMs, starting with ChatGPT. What is an LLM like ChatGPT? It is essential to distinguish between two primary operations when it comes to models like ChatGPT: training and inference. During training, vast amounts of data, often encompassing vast swaths of the internet, are fed into the model. This training data enables the model to represent and recognize patterns, contexts, and relationships in human language. The entire training process is intricate and involves steps such as pre-training, supervised fine-tuning, reward modeling, and reinforcement learning.[83] Techniques such as tokenization help break down language into digestible bits for the model. It is noteworthy that training these models is not cheap, with costs spanning computational infrastructure, data storage, and energy consumption.

---

[83] Shervin Minaee, T. M., Narjes Nikzad, Meysam Chenaghlu Richard Socher, Xavier Amatriain, Jianfeng Gao. (2024). Large Language Models: A Survey. *Working Paper*. https://arxiv.org/pdf/2402.06196

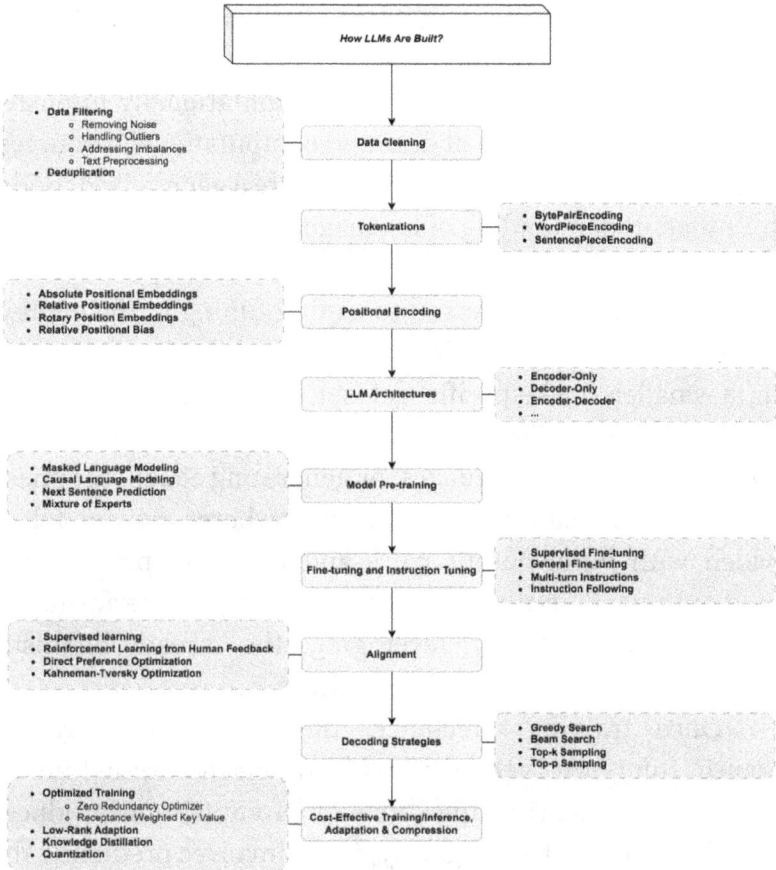

At its essence, an LLM is a model of a vast amount of human thought, manifested and interlinked through language. Before training, the text of the internet is encoded in tokens—concepts, thoughts, and chains of thoughts—that can be learned and associated with other tokens. Pre-training involves training the model on a very large amount of text data, usually from the internet, books, and other extensive text sources. During pre-training, the model learns to predict the next word in a sentence, capture syntactic and semantic relationships, and develop an understanding of language patterns. In this sense, LLMs are simply extending the predictive power of AI into new domains with new models.[84] [85] This phase is computationally intensive and requires vast amounts of data and computational resources. The primary goal of pre-training is to create a generalized model with a broad understanding of language, which can then be fine-tuned for specific tasks.

Training, often referred to as fine-tuning in the context of LLMs, takes the pre-trained model and further adjusts it using a smaller, task-specific dataset. Fine-tuning allows the model to specialize in particular tasks such as answering questions, translating languages, or generating code. This phase usually involves supervised learning, where the model is provided with input-output pairs and learns to produce the desired outputs. Fine-tuning enhances the model's performance in specific applications, leveraging the broad linguistic knowledge acquired during pre-training.

Once training concludes, the model is ready for inference. Inference is the process in which the trained model takes in new inputs (like questions or prompts) and produces outputs based on its training. It typically involves predicting the next word in a sequence, which, when done repeatedly, generates complete responses. The accuracy and coherence of these outputs underscore the sheer volume of information and pattern recognition the model has internalized, and its capacity

---

[84] Agrawal, A., Gans, J., & Goldfarb, A. (2018). *Prediction machines: the simple economics of artificial intelligence.* Harvard Business Review Press.
[85] Bernd Carsten Stahl, J. A., Nitika Bhalla, Laurence Brooks, Philip Jansen, Blerta Lindqvist, Alexey Kirichenko, Samuel Marchal, Rowena Rodrigues, Nicole Santiago, Zuzanna Warso, David Wright (2023). A systematic review of artificial intelligence impact assessments. *Artificial Intelligence Review, 56,* 12799-12831.

to infer an appropriate response to the prompt based on the model.

The beating heart of GPT and similar models is the "transformer" architecture, epitomized by an influential paper titled "Attention is All You Need", produced by researchers at Google. This architecture made significant improvements in processing sequences in data, especially language. While the term "deep learning" often floats around in AI discussions, it is a category of machine learning that uses neural networks with multiple layers. The transformer architecture is a type of deep learning model, allowing LLMs to process and generate human-like text.

One cannot discuss LLMs without touching on "prompting." Prompts are initial inputs or questions given to the model, guiding the AI to produce a specific type of output. The AI's response can be as succinct as a single word or as extensive as an entire essay. It infers what is the best response to the prompt based on its model of all prior language content in its training data and the interconnections between them.

To appreciate the versatility and breadth of LLM applications, let's compare some models. DALL·E, for instance, is a derivative of GPT-3 that turns textual descriptions into images. The success of DALL·E reinforces a fascinating notion: language, particularly English, has evolved into an extraordinarily versatile coding language. Beyond GPT and DALL·E, the underlying technological stack includes infrastructural tools from companies like NVIDIA, foundational models like ChatGPT, toolkits offered by entities like Hugging Face, and specialized applications such as Harvey for legal text generation.

At the heart of these technologies lie several core functionalities:

- Prompting: The art of effectively guiding AI to produce desired outputs.
- Plugins/Apps: Tailored applications built on top of foundational AI models to serve specific industries or tasks.
- Ecosystem tools: These are supporting tools and platforms that facilitate the development, fine-tuning, and deployment of AI models.

Lastly, a distinction worth highlighting is between foundational models and fine-tuned models. Foundational models are the broad, general models trained on extensive data. They are like Jacks-of-all-trades, trained on vast amounts of data in many categories, and can effectively make inferences in many domains. In contrast, fine-tuned models are specialized versions of these foundational models, tailored for specific tasks or industries, making them masters of limited domains.

In conclusion, the landscape of generative AI, especially LLMs, is vast, intricate, and profoundly transformative. The confluence of historical research, technological advancements, and innovative applications has brought forth an era where machines don't just predict—they generate new creations.

An Intelligence Framework for Generative AI

The technical advancement fueling the modern surge in generative AI was the encapsulation of human language within machine learning models that could be used to produce fresh inferences based on simple prompts—that is, the introduction of OpenAI's ChatGPT LLM, which enabled millions of users to access this powerful language model. Human language is inherently multifaceted, encoding not merely semantic meaning but also rhetoric, conversation, debate, planning, inventiveness, and on-the-fly reasoning. Language and human cognition are so intimately entwined that some cognitive scientists propose that the advent of sophisticated thought and human intelligence itself was reliant on language's evolution.

Yet, it is pivotal to understand that the term "intelligence" encapsulates a myriad of capabilities. Cognitive scientists sometimes bifurcate these capabilities, labeling them as different "intelligences." Essentially, intelligence comprises a suite of capabilities to engage with the world—this includes perception, comprehension, problem-solving, and informed action-taking. It is almost axiomatic that individuals possessing higher intelligence tend to excel in most aspects of life, spanning academics, professional pursuits, financial achievements, health, leadership roles, entrepreneurial endeavors, artistic ventures, and even overall life satisfaction.

As Marc Andreessen has argued, intelligence has been the lever that has created the world we live in today, having raised our standard of living dramatically. Andreessen perceives AI as a potent catalyst, poised to significantly augment human intelligence across countless domains, enhancing all intelligence-driven outcomes. What is difficult to predict, however, is exactly what areas of life AI will enhance and at what rate. Indeed, the augmentation of different tasks could occur at different rates,[86] thus thwarting those who would seek to

---

[86] Dell'Acqua, F., McFowland, E., Mollick, E. R., Lifshitz-Assaf, H., Kellogg, K., & Lakhani, K. R. (2023). *Navigating the jagged technological frontier: field experimental evidence of the effects of AI on knowledge worker productivity and quality* (Harvard Business School Technology & Operations Mgt. Unit Working Paper, Issue.

automate complete activities.[87] Every groundbreaking technology possesses unique capabilities—or "affordances" as scholars in information technology would call them—that transcend its predecessors. Affordances are the particular utilities a technology facilitates. Interestingly, these affordances often are not immediately evident in marketing or technical manuals. Instead, they often emerge organically, unearthed by users over time.

Affordances of Artificial Intelligence: Conceptual Tasks with Language

The gamut of affordances ushered in by generative AI is only now becoming clear. Within the broad spectrum representing intelligence, LLMs particularly shine in conceptual tasks where language is the central medium. The key is the generation of content based on an inference from a prompt. Yet this masks the specific affordances that are possible from this technology, including how conversational or multi-model it can be.[88] Although LLMs can potentially branch out from purely conceptual tasks—as exemplified by DALL·E, demonstrating that LLM's language representation might be an apt foundation for generating images—their primary domain currently revolves around language-related conceptual tasks. Thus, when AI with LLMs supplements human intelligence, it predominantly enhances the execution and results of specific conceptual undertakings.

---

[87] Raisch, S., & Krakowski, S. (2021). Artificial intelligence and management: The automation–augmentation paradox. *Academy of Management Review*, *46*(1), 192-210.

[88] Glaser, V. L., & Gehman, J. (2023). Chatty Actors: Generative AI and the Reassembly of Agency in Qualitative Research. *Journal of Management Inquiry*.

The landscape of LLMs is rapidly evolving, driven by substantial investments by major tech companies.[89] OpenAI's ChatGPT, Google DeepMind's Gemini, Anthropic's Claude, Microsoft's Copilot, and Perplexity AI's Perplexity are some of the leading LLMs currently shaping the AI ecosystem. ChatGPT, developed by OpenAI and launched in 2020, leverages the GPT-3 and subsequent GPT-4 architectures, known for their expansive datasets and deep learning capabilities. Google DeepMind's Gemini, showcases Google's commitment to AI, integrating cutting-edge transformer models to enhance natural language understanding and generation. On the other hand, Anthropic's Claude, named after Claude Shannon, emphasizes safety and alignment in AI applications, setting it apart with its unique focus on ethical AI use. Microsoft's Copilot, embedded within its developer tools, utilizes the Codex model to assist in code generation and software development, highlighting Microsoft's strategy to integrate AI into productivity tools. Meanwhile, Perplexity, developed by Perplexity AI, is known for its question-answering capabilities and conversational AI, leveraging advanced natural language processing techniques.

The models have also been rated for their capabilities in different knowledge domains. For example, some have said that Gemini is particularly adept at finance, Copilot excels at creative writing, Perplexity is best at summarization of citations, and ChatGPT is good at health-related features.[90] Each of these models reflects its creators' strategic priorities and technological

[89] Naveed, H., Khan, A. U., Qiu, S., Saqib, M., Anwar, S., & Mian, A. (2023). A comprehensive overview of large language models. *Working Paper.* https://arxiv.org/abs/2307.06435

[90] Brown, D., Dapena, K., & Stern, J. (2024). The Great AI Challenge: We Test Which Bot Is Best. *Wall Street Journal.* https://archive.is/2024.05.25-094849/https://www.wsj.com/tech/personal-tech/ai-chatbots-chatgpt-gemini-copilot-perplexity-claude-f9e40d26

advancements, as well as their unique training datasets, showcasing diverse approaches and investments in AI.

	FIRST	SECOND	THIRD	FOURTH	FIFTH
OVERALL	Perplexity	ChatGPT	Gemini	Claude	Copilot
HEALTH	ChatGPT	Gemini	Perplexity	Claude	Copilot
FINANCE	Gemini	Claude	Perplexity	ChatGPT	Copilot
COOKING	ChatGPT	Gemini	Perplexity	Claude	Copilot
WORK WRITING	Claude	Perplexity	Gemini	ChatGPT	Copilot
CREATIVE WRITING	Copilot	Claude	Perplexity	Gemini	ChatGPT
SUMMARIZATION	Perplexity	Copilot	ChatGPT	Claude	Gemini
CURRENT EVENTS	Perplexity	ChatGPT	Copilot	Claude	Gemini
CODING	Perplexity	ChatGPT	Gemini	Claude	Copilot
SPEED	ChatGPT	Gemini	Copilot	Claude	Perplexity

The differences between major LLMs such as ChatGPT, Gemini, Claude, Copilot, and Perplexity can be attributed to their varying technical specifications, including model size, architecture, and capabilities. Model size, typically measured in the number of parameters, directly influences an LLM's ability to generate coherent and contextually relevant responses. For instance, OpenAI's GPT-4, which powers ChatGPT, boasts hundreds of billions of parameters, allowing for nuanced and sophisticated language generation. In contrast, models like

Copilot, tailored for specific tasks such as code generation, may optimize their architecture to balance performance and efficiency. The capabilities of these models are also shaped by their training data and objectives—for example, Claude focuses on AI safety and ethical considerations, influencing its response generation differently than a productivity-oriented model like Copilot. The Elo score, traditionally used in chess to rate players' relative skill levels, is now applied in AI to assess model performance. An LLM's Elo score is determined through comparative evaluations against other models, providing a benchmark for its language understanding and generation proficiency.[91]

---

[91] Chang, Y., Wang, X., Wang, J., Wu, Y., Zhu, K., Chen, H., & Xie, X. (2023). A survey on evaluation of large language models. *Working Paper*. https://arxiv.org/abs/2307.03109

🏆 LMSYS Chatbot Arena Leaderboard

Vote!

Blog | GitHub | Paper | Dataset | Twitter | Discord | Kaggle Competition

LMSYS Chatbot Arena is a crowdsourced open platform for LLM evals. We've collected over 1,000,000 human pairwise comparisons to rank LLMs with the Bradley-Terry model and display the model ratings in Elo-scale. You can find more details in our paper. **Chatbot arena is dependent on community participation, please contribute by casting your vote!**

Arena   NEW: Arena (Vision)   Full Leaderboard

Total #models: **116.** Total #votes: **1,498,403.** Last updated: 2024-07-16.

NEW! View leaderboard for different categories (e.g., coding, long user query)! This is still in preview and subject to change.

Code to recreate leaderboard tables and plots in this notebook. You can contribute your vote at chat.lmsys.org!

Category   Overall Questions

Overall   #models: 116 (100%)   #votes: 1,498,403 (100%)

Rank* (UB)	Model	Arena Score	95% CI	Votes	Organization	License	Knowledge Cutoff
1	GPT-4o-2024-05-13	1287	+3/-4	62251	OpenAI	Proprietary	2023/10
2	Claude 3.5 Sonnet	1271	+3/-3	31482	Anthropic	Proprietary	2024/4
2	Gemini-Advanced-0514	1267	+4/-4	46549	Google	Proprietary	Online
3	Gemini-1.5-Pro-API-0514	1262	+3/-3	55037	Google	Proprietary	2023/11
4	Gemini-1.5-Pro-API-0409-Preview	1257	+4/-3	55678	Google	Proprietary	2023/11
4	GPT-4-Turbo-2024-04-09	1257	+3/-3	75215	OpenAI	Proprietary	2023/12
7	GPT-4-1106-preview	1251	+3/-3	87739	OpenAI	Proprietary	2023/4
7	Claude 3 Opus	1248	+2/-2	146594	Anthropic	Proprietary	2023/8
7	GPT-4-0125-preview	1245	+3/-2	81062	OpenAI	Proprietary	2023/12
9	Yi-Large-preview	1240	+3/-3	49814	01 AI	Proprietary	Unknown
11	Gemini-1.5-Flash-API-0514	1228	+4/-3	46614	Google	Proprietary	2023/11

These Elo scores are calculated based on over one million human pairwise comparison votes. As the different big tech companies are often improving models and releasing new ones, their relative performance can be tracked over time.[92] Of course, these comparative ratings are likely to change over time, so it is important to check the current ratings before using them for a particular task.

---

[92] Chiefaioffice. (2024). *Foundational model wars over the past 12 months (Elo Scores by Company)*.
https://x.com/chiefaioffice/status/1793407809847275864

Elo Scores by Company
Top Ranked Model by Company in the Chatbot Arena

Replay

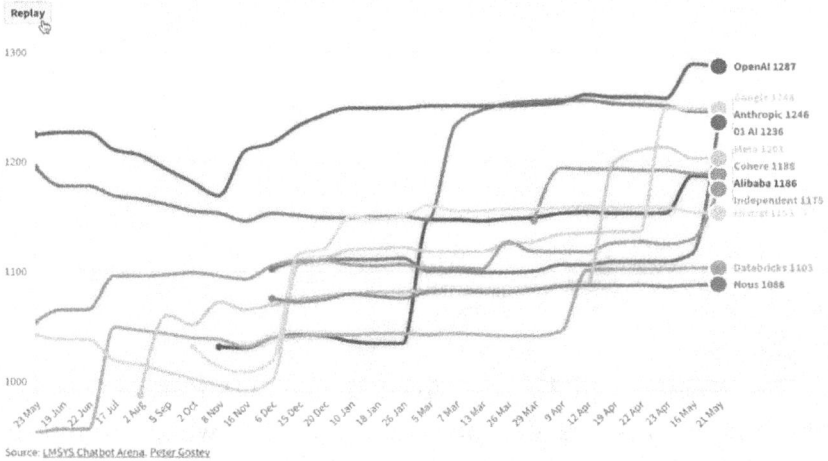

Source: LMSYS Chatbot Arena, Peter Gostev

These technical differences and performance metrics highlight the diverse strengths and specializations of each LLM, reflecting their creators' strategic objectives and investment priorities.

Human Reasoning with Chain-of-thought Inference

In one interesting development, OpenAI's new model series, codenamed "Strawberry" or the "O1" models, including the "O1 Preview" and "O1 Mini," represent a significant leap in generative AI's ability to handle complex reasoning tasks. A key feature of these models is their chain-of-thought processing at the point of inference, which involves generating and utilizing many hidden tokens to simulate step-by-step reasoning. This enables them to handle tasks that require human-like reasoning, such as solving mathematical problems, making scientific deductions, and planning solutions for complex logical challenges. The "O1" models' strength in these areas has led many to liken their performance to that of an incoming PhD student.[93]

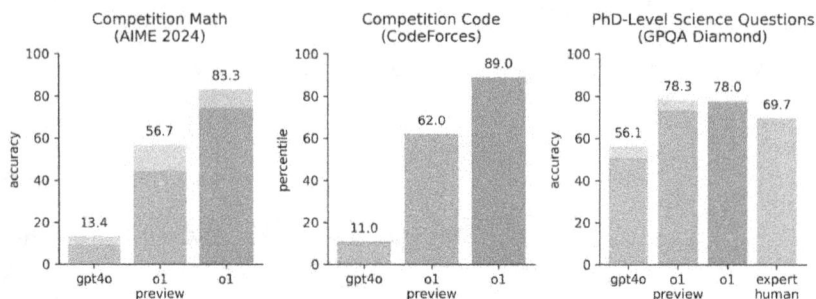

Competition Math (AIME 2024)	Competition Code (CodeForces)	PhD-Level Science Questions (GPQA Diamond)

Notably, an IQ test given to the "O1 Preview" model resulted in a score of approximately 120, which compares favorably to prior models like ChatGPT-4, LLAMA-3.1, and Claude-3 Opus, which typically scored between 60 and 90.[94] This boost in cognitive capabilities comes at the cost of significantly more computational resources at the inference stage, as the model's chain of thought processes require intensive calculation. Looking forward, this trend could favor companies with access to specialized inference hardware and cloud providers that can offer the substantial computational power needed to fully leverage these advancements.

Major tech players have been quick in elucidating what they discern as the prime affordances of generative AI. NVIDIA, for instance, showcases a suite of applications LLM facilitates for businesses, emphasizing cognitive capabilities. Likewise, OpenAI, alongside Andrew Ng of DeepLearning.AI, has rolled

[93] OpenAI. (2024). *Learning to Reason with LLMs.* September 12, 2024 https://openai.com/index/learning-to-reason-with-llms/

[94] Lott, M. (2024). *Massive breakthrough in AI intelligence: OpenAI passes IQ 120.* September 14, 2024 https://www.maximumtruth.org/p/massive-breakthrough-in-ai-intelligence

out resources detailing LLM applications focused on diverse information manipulations. Some scholars have explored this space as well, with noteworthy contributions like the extensive compilation of LLM use cases under "cognitive automation" by University of Virginia's Professor Anton Korinek.[95] Drawing from these varied sources and my own research, I present an integrated and expanded classification centered on three conceptual task categories: 1) information integration, 2) predictive inference, and 3) knowledge transformation. This represents an intelligence framework for understanding the conceptual tasks underlying generative AI. A summary of the Generative AI Intelligence Framework—including the conceptual tasks and their impact—appears in Table 2.

---

[95] Korinek, A. (2023a). Generative AI for Economic Research: Use Cases and Implications for Economists. *Journal of Economic Literature, 61*(4), 1281-1317.

Table 2: Generative AI Intelligence Framework: Conceptual Tasks and Impact

Conceptual Task Categories	Information Integration	Predictive Inference	Knowledge Transformation
Core Function	Gathering, organizing, and presenting data to facilitate decision-making and understanding.	Drawing logical conclusions from data to predict trends and outcomes.	Modifying and enhancing information to generate new insights or tailor them to specific needs.
Subtasks	Summaries, search, integration, comparison	Classification, sentiment analysis, topic extraction, fraud analysis, proof of human	Translation, in-depth comparison, copyediting, expansion, performativity ("act as"), customization/ personalization, data analysis/ inference
Examples of Use	Automating literature reviews, real-time information retrieval, merging insights from reports.	Detecting customer sentiment in social media, classifying emails, fraud detection in financial transactions.	Translating documents, expanding summaries into detailed reports, customizing content for different audiences.
Output Detail	Highly structured outputs like summaries,	Structured outputs like classifications, sentiment	Highly detailed outputs, often involving creative or analytical

	integrated report s, and comparative analyses.	scores, topics, or fraud alerts.	processing, such as translations or expanded texts.
User Involveme nt	Low to medium; most tasks require simple inputs from users and yield direct outputs.	Medium to high; requires more user interaction to set criteria and validate predictions.	High; tasks often require user-specific inputs and feedback loops to ensure accuracy and relevance.
Complexity of Implement ation	Moderate; requires effective data parsing and organization algo rithms but is relatively straightforward.	High; involves complex algorithms to detect patterns and anomalies, requ iring robust machine learning models.	Very high; involves complex processing to adapt, transform, and generate content that meets specific user needs.
Potential Business Impact	High; enhances decision-making speed and accuracy, making information processing more efficient.	Very high; can significantly improve operational effi ciency, reduce fraud, and enhance customer understanding.	Extremely high; critical for businesses needing tailored content, global reach, and nuanced understanding of data.

Intelligence Framework: Conceptual Tasks where LLMs Excel

The first category of conceptual tasks where LLMs excel is around information integration. Information integration is the capacity to access, organize, and present data in a manner that optimally serves as an input for the user's specific tasks. Prominent examples of information integration capabilities are below.

Information Integration

Summaries. LLMs like ChatGPT can condense vast amounts of textual information into brief yet informative summaries. This capability allows users to quickly grasp the essence of extensive documents, news articles, or research papers, making it easier to digest substantial information in shorter time frames.

Search. Utilizing LLMs transforms the search process. By fetching pertinent information from extensive datasets or the internet in real-time, ChatGPT ensures users are presented with accurate and contextually relevant results. This adaptability makes information retrieval both faster and more accurate.

Integration. LLMs excel at weaving together data from varied sources, offering users an integrated and cohesive viewpoint. For instance, when exploring a topic, they can merge insights from disparate articles, ensuring users receive a synthesized understanding, eliminating the need to cross-reference multiple sources.

Comparison. When users need to contrast pieces of information, LLMs provide a structured comparative analysis. By highlighting differences and similarities, these models offer a comprehensive understanding, crucial for informed decision-making.

Predictive Inference

The second category is what I call predictive inference. Predictive inference pertains to the ability to draw logical conclusions from data, anchored on projections that enhance the odds of achieving user-set objectives. Prominent examples of predictive inference are below.

Classification. LLMs efficiently categorize diverse pieces of information. They can sort data based on user-specified criteria, be it segregating types of emails, differentiating customer feedback, or classifying product genres. Such categorization streamlines processes and simplifies information navigation.

Sentiment analysis. LLMs assess the emotional tone of text. By detecting sentiment, they enable businesses to gauge how customers feel about products or services, giving them insights into areas of improvement or success. In essence, they transform text into actionable emotional feedback.

Topic extraction. By sifting through vast content, LLMs can identify and highlight main themes, ensuring users can instantly comprehend the core subjects of extensive texts without the need to read them in their entirety.

Fraud analysis. These models, trained to spot irregular patterns in datasets, are instrumental in detecting fraudulent activities. They enhance security by identifying potential threats or anomalies, particularly in sectors such as finance.

Proof of human. To counteract automated bots, LLMs can validate genuine human interactions on online platforms.

By generating and evaluating specific interaction patterns, they ensure that users are authentic.

Knowledge Transformation

The third category is knowledge transformation. Knowledge Transformation encompasses the ability to modify, augment, and morph information into user-specified knowledge. Prominent examples of knowledge transformation are below.

Translation. Language is no barrier for LLMs. Equipped with extensive linguistic databases, they provide seamless translations across a myriad of languages, bridging communication gaps and facilitating global interactions.

Comparison. In knowledge transformation, LLMs do not merely integrate. They offer in-depth insights and rich contrasts between complex concepts, promoting a deeper understanding and more nuanced perspectives.

Copyediting. Beyond mere textual generation, LLMs refine and perfect written content. They ensure texts are free from spelling or grammatical errors, thereby enhancing clarity and coherence.

Expansion. When presented with concise information, LLMs can extrapolate and provide comprehensive elaborations. For example, if given a brief statement about a historical event, they could generate a more detailed explanation or backstory.

Performativity ("act as"). LLMs can simulate specific roles or personas during interactions, catering their responses to fit roles like a tech guru, a history buff, or a financial consultant. This versatility enhances user interactions, providing targeted and relevant information.

Customization/ Personalization. Recognizing user preferences and past interactions, LLMs can mold information delivery, ensuring it remains pertinent and engaging for the user, enhancing user experience through tailored interactions.

Data Analysis/ Inference. LLMs go beyond data presentation. They analyze data, drawing out patterns, trends, and conclusions, ensuring users are equipped with insights that drive informed decision-making.

LLM Product Architectures: Emerging Dominant Designs

It is crucial to understand that the LLM is not just a standalone product; it serves as the technology foundation for different product types. OpenAI's ChatGPT, embodied in a straightforward website and app with a prompting text box, is the simplest possible product architecture. Such a basic configuration could potentially unlock the vast capabilities of an LLM for consumers. It is not surprising that other companies have copied this simple LLM product architecture. The rapid adoption of this minimalistic platform is quite remarkable. Unlike platforms like YouTube and TikTok, which largely offer passive content consumption apart from clicking and swiping, ChatGPT demands active user engagement in every iteration to generate fresh content. Such user-intensive popular products have not been seen since the dawn of search engines.

Nevertheless, the initial success of ChatGPT should serve as a marker for the array of product architectures that could evolve with LLMs as their foundational technology. Just as the microprocessor gave rise to desktops, laptops, smartphones, tablets, and a wide range of other devices that serve different customer needs, LLM technology will give rise to many other product architectures. These range in complexity, cater to different target audiences, and span various applications. The exact future trajectory of LLM-based generative AI product architectures remains a mystery. However, certain frameworks are taking shape. Each of these is vying to become the "dominant design" in its product category. A dominant design is a product iteration that gains significant traction in the market, compelling competitors and new entrants to adopt a similar framework to grow market share. The rise of such dominant designs propels the explosive growth seen in the S-curve model of innovation, as they provide a unified direction for innovators to channel their investments and efforts as product iterations improve dramatically.

Even at this nascent stage of LLM evolution, three distinct product architectures that show promise in becoming dominant designs can be identified. My analysis focuses on these three that are aiming to be General Purpose Technologies for use by both consumers and enterprises. (It is worth noting that there are numerous other specialized product architectures, including those that are tailored to utilize data from singular enterprises.) These three emerging architectures vary significantly in user interaction levels, autonomy, and the range of conceptual tasks they can perform. The crux lies in understanding each architecture by dissecting how it undertakes its conceptual tasks and to what end. Below, I describe these three emerging LLM product architectures in detail.

These emerging architectures extend beyond mere chatting functionality. They differ in the level of user interaction and autonomy, the integration with operating workflows, and the variety and modalities of conceptual tasks they perform. Each architecture can be analyzed in terms of how it accomplishes its tasks and to what end. The three notable architectures are:

Content Generators

The genesis of LLM technology's application can be traced to what I term "content generators." This includes pioneering platforms like DALL-E and ChatGPT, which started as simple websites and apps capable of generating content from straightforward text-based prompts. This architecture was a significant leap forward, as it laid bare the immense potential of massive language models for mainstream users. It opened a limitless arena of prompts, inviting users to delve into this expanse for various productive applications. Following these frontrunners, other companies soon entered the fray, producing their own content generators like Midjourney in image generation and a slew of language generator products such as Bing, Gemini, LLaMA, and Anthropic.

The predominance of the content generation architecture as a standard design is evident not only in its widespread adoption but also in the flurry of incremental innovations that have ensued since its establishment. These innovations have aimed to refine inference quality and introduce new prompt modalities, including images and voice. Despite these advancements, content generators are not without their drawbacks. They require continuous user input and adjustments, which may not always be ideal. As a result, they often work best as an extended "chat" where the user gives constant direction and refinement. However, their level of interactivity falls short when compared to more dynamic interactions, such as those with a skilled teacher, in that they do not engage in ongoing dialogue after providing their responses. In other words, they respond but do not initiate. Additionally, the complexity of problems they can address is often limited to simpler tasks, although with constant iteration and refinement, content generators can be used to generate more complex outputs.

Copilot Assistants

Another significant product architecture in the LLM space is the AI-based assistant, often referred to as a "copilot." Early manifestations of LLM-based copilots appeared in the realm of computer programming, notably with GitHub's Copilot. This tool, trained on a vast repository of code samples in the GitHub repository, suggests new code snippets, effectively turning the programmer's existing code into a prompt for further development. This form of copilot offers semi-autonomy, and is able to generate coding suggestions proactively based on the user's ongoing input. Another example is Replit's programming copilot, integrated into its development environment.

The versatility of the copilot concept means it can be adapted to virtually any workflow, a fact not lost on major tech companies. For instance, Google incorporated a "help me write" function in Google Docs. However, it was Microsoft's introduction of "Copilot" into its Office suite that truly mainstreamed the assistant paradigm in individual productivity software. Microsoft's Copilot extends across various Office products, offering specialized assistance in tasks like writing in Word, slide development in PowerPoint, email drafting in Outlook, and data analysis in Excel. Powered by generative AI and custom LLMs, these copilots significantly enhance and accelerate the workflows of knowledge workers. There is some evidence that organizations are struggling to integrate copilot into its workflows, perhaps because it is so difficult to change organizational routines and policies.[96] This may take time to adapt. The key is that copilot assistants "work alongside" users in well-defined workflows where they can suggest and facilitate improvements in a structured manner.

Autonomous Agents

The most advanced and still-evolving product architecture within the LLM space is that of autonomous agents. These systems are designed to pursue complex objectives with minimal human oversight. They are agents in that they proactively execute actions within their capability to achieve set goals. For example, an autonomous agent tasked with building a website about academic activities would independently gather research papers, teaching history, and relevant background information, create and select suitable images, write webpage content, and even handle the HTML coding. In more intricate scenarios, various agents might interact, such as a stock trading agent collaborating with agents specializing in company press-release analysis, market pricing and technical analysis, and monitoring financial social media, integrating these insights to formulate a strategy.

Although still in the early stages of development, autonomous agents have already found applications in diverse domains. Many leverage existing LLMs like ChatGPT, DALL-E, or open-source models like Mistral, integrating these via APIs. Platforms like Microsoft's Autogen system are emerging to facilitate the creation of new autonomous or multi-agent systems. Autogen Studio provides a graphical interface for managing these agents and their interplay centrally. In contrast,

---

[96] Benioff, M. (2024). *75% said their employees were struggling to integrate Microsoft Copilot into their daily routines.*
https://x.com/Benioff/status/1858314876864598170

Autonolas offers a blockchain protocol for decentralized management and ownership of autonomous agents, with numerous applications already active in financial services. These developments underscore the diverse potential and evolving nature of autonomous agent systems in the LLM landscape. The startup company Replit recently announced an autonomous agent programming tool that fully constructs new apps in Replit's IDE—although other LLM coding packages exist, this one was notable for its autonomy going from natural language to full apps quickly.[97]

A final addition is that OpenAI has announced that it will support AI agents of two types—those that "operate devices" and "automate tasks", although what form this will take is unclear.[98] This may signal a move from LLMs to "large action models" or "large agent models" (LAMs).[99] Not to be outdone, Mistral has also incorporated an agent's feature, which enables users to customize and deploy agents for free and after only minutes of use.[100] We should expect many LLM vendors to offer agent-based systems.

A broader advance in online autonomy may be systems that can operate as we do online. Indeed, OpenAI is launching new software codenamed "Operator" that can take control of

---

[97] Masad, A. (2024). *Announcing Replit Agent.* Replit. September 6, 2024

[98] Palazzolo, S., & Efarti, A. (2024). OpenAI Shifts AI Battleground to Software that Operates Devices, Automates Tasks. *The Information.* https://www.theinformation.com/articles/openai-shifts-ai-battleground-to-software-that-operates-devices-automates-tasks?utm_source=ti_app

[99] Dobos, N. (2024). *2004 is the year of Large Action Models.* https://x.com/NickADobos/status/1755355930181652494

[100] Alvaro, C. (2024). *Mistral's new agents feature.* 2024-08-18 https://x.com/dr_cintas/status/1825182614170263910

computing resources like browsers to perform tasks like finding information, clicking buttons, and installing software.[101]

The challenges of using agents mainly stem from known challenges in the engineering of goal-directed and coupled systems. The promise of autonomous agents is that they can flexibly pursue goals without intervention.[102] Where one agent may be insufficient, multiple agents can work in parallel or in coordination—indeed, some research indicates that more agents help improve accuracy on various tasks.[103] However, the key challenge is that agents can become correlated and too similar in their responses, in turn generating outputs that are too similar, to create additional improvements. Another issue is that coupling agents together can magnify errors. As the team at Parcha, a multi-agent startup, found, "If an AI agent carries out a workflow consisting of 10 tasks autonomously but has a 10% error rate per task, the compounded error rate over the whole workflow is 65%," leading to catastrophic failure.[104] Multi-agent systems are a topic of great interest where research is likely to advance quickly. A final issue is that highly autonomous agents may be less trusted if they are found to

---

[101] Ghaffary, S., & Metz, R. (2024). OpenAI Nears Launch of AI Agent Tool to Automate Tasks for Users. *Bloomberg*. https://www.bloomberg.com/news/articles/2024-11-13/openai-nears-launch-of-ai-agents-to-automate-tasks-for-users

[102] Sato, M. K., Koba, L. J., Du, H., Goodrich, B., Hasin, M., Chan, L., Miles, L. H., Lin, T. R., Wijk, H., Burget, J., Ho, A., Barnes, E., & Christiano, P. (2023). Evaluating Language-Model Agents on Realistic Autonomous Tasks. In.

[103] Li, J., Zhang, Q., Yu, Y., Fu, Q., & Ye, D. (2024). More Agents Is All You Need. *Working Paper.* https://arxiv.org/abs/2402.05120

[104] Berrios, M. R. (2024). *Lessons from Parcha's Journey automating compliance workflows using AI and why autonomous agents aren't always the best solution.* Parcha's Resources. 2024-06-06 https://guidetoai.parcha.com/agents-arent-all-you-need/

betray that trust.[105] Thus, autonomous systems need to ensure they match expectations of user outputs.

In summary, these three architectures—content generators, copilot assistants, and autonomous agents—vary in key aspects such as autonomy, user interaction level, workflow integration, and modality range. From one-shot interactions in content generators to fully autonomous operations in agent-based systems, these architectures are shaping the landscape of LLM product design and usage. These product architectures are summarized in Table 3.

---

[105] Vanneste, B. S., & Puranam, P. (2024). Artificial Intelligence, trust, and perceptions of agency. *Academy of Management Review.* ttps://doi.org/10.5465/amr.2022.0041

Table 3: Generative AI Product Architectures

Dimension	Content Generators	Copilot Assistants	Autonomous Agents
Description	Platforms that generate content based on user-provided prompts. These range from text generation tools like ChatGPT to image generation platforms like DALL-E. They are primarily used for creating content directly from user input.	Semi-autonomous tools designed to assist users within specific workflows. They generate suggestions or perform tasks in parallel with users, improving efficiency and accuracy.	Fully autonomous systems capable of executing complex tasks with minimal human intervention. They work towards specific goals by making decisions and taking actions independently.
Examples of Use	Generating articles, blog posts, images, or even audio files from simple text prompts. Used in creative industries, marketing, and for personal productivity tasks.	Programming assistants like GitHub Copilot, writing aids in Google Docs, and productivity enhancements in Microsoft Office, such as Excel or Word Copilot.	Automating multi-step processes like web scraping, stock trading, or even creating and managing websites. Used in areas like finance, research, and digital marketing.
Prompting Patterns	Direct prompts from users initiate content creation. Some platforms offer back-and-forth interaction, but	User prompts drive the assistant's actions, but workflow-generated content also acts	Initial prompts set goals, but agents generate subsequent prompts from their own actions and outputs,

	content generation usually involves a series of individual prompts.	as implicit prompts, guiding the assistant's suggestions.	operating in a continuous cycle till objectives are met.
Autonomy Level	Low to medium; content generators rely heavily on user input and guidance at each step.	Medium; operates alongside users, offering suggestions and completing tasks with some autonomy.	High; designed to function independently, requiring little to no human guidance after initial setup.
User Interaction Level	High; requires constant user input to guide content creation and make refinements.	Medium to high; interaction varies by task, with users needing to refine or approve assistant-generated outputs.	Low; user interaction is limited to setting initial goals or parameters.
Integration with Workflows	Minimal integration with workflows; typically operate as standalone applications.	Highly integrated with workflows; designed to complement an d enhance specific user tasks within established processes.	Deep integration with workflows, often taking over entire processes and working across multiple systems.
Complexity of Tasks	Low to medium; best suited for simpler tasks	Medium to high; capable of handling more	High; capable of handling highly complex, multi-

	like generating short content or images, though iterative input can yield more complex outputs.	complex tasks like coding, data analysis, and document drafting.	step tasks that require ongoing analysis, decision-making, and execution.

Prompting: Best Practices for Inference from LLMs

Despite the diversity of product architectures that use LLM technology, including the three emerging ones described above and others yet to be discovered, they all include one fundamental interface for taking in customer inputs: prompting. In content generators, prompts are the main interface, directly shaping the generation of text, images, or audio. Copilot assistants also rely on user prompts, but they incorporate content produced within the workflow as well—such as code, text, or slides—as additional input sources. This workflow-generated content acts as an implicit prompt, guiding the LLM to produce further suggestions. Autonomous agents, meanwhile, can accept user-generated prompts as well, but they also use the results of their own activities, like web searches or analyses, as prompts for subsequent tasks. These product architectures are depicted in Figure 3.

Figure 3: Generative AI Product Architectures: Human and Computer Interactions over time

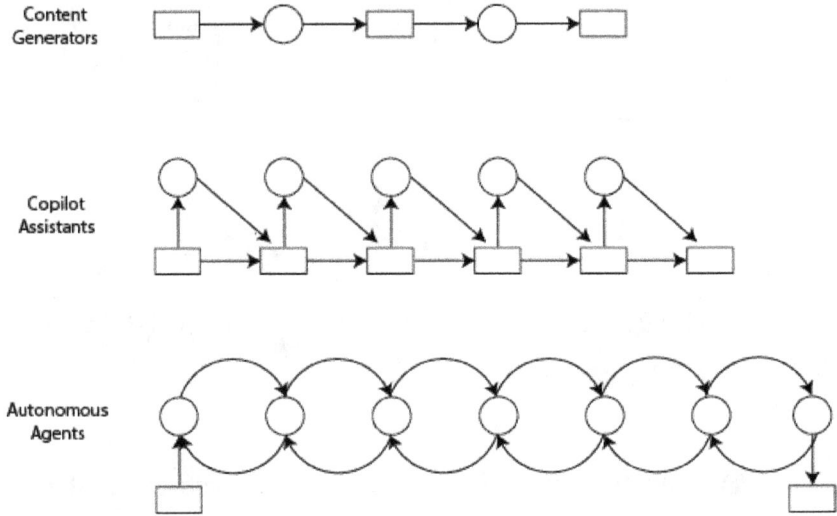

Understanding effective prompts are crafted is crucial in eliciting the desired output from an LLM. The challenge many face is in engineering the right prompts to achieve satisfactory results. The initial awe of using ChatGPT can quickly turn to disappointment when simple prompts fail to produce complex, subtle, or appropriate outcomes. The key to unlocking significant productivity gains lies in grasping the emerging best practices of prompting. Fortunately, there is a significant literature emerging on effective prompting techniques to draw upon here as indicated below.[106]

---

[106] Reddy, B. (2024). *The Power of Prompt Engineering.* July 20, 2024
https://x.com/bindureddy/status/1814409737557160044

NLP Tasks		
	Mathematical Problem Solving	CoT, Random CoT, Complex CoT, Basic, PAL, Synthetic Prompting, Contrastive CoT, Contrastive Self-Consistency, CoC, Auto-CoT, Self-Consistency, Active-Prompt, PS, PoT, MathPrompter, ToT, LoT, Fed-SP-SC, Fed-DP-CoT, Analogical Reasoning, Least-to-Most [Yasunaga et al. (2023), Wei et al. (2022), Zhang et al. (2022), Wang et al. (2022), Yao et al. (2024), Zhao et al. (2023b), Chen et al. (2022a), Li et al. (2023a), Gao et al. (2023), Liu et al. (2023), Chia et al. (2023), Diao et al. (2023), Shao et al. (2023), Zhou et al. (2022), Imani et al. (2023), Fu et al. (2022), Wang et al. (2023)]
	Logical Reasoning	Basic, CoT, PAL, Synthetic Prompting, CoC, LoT, ToT, Analogical Reasoning [Yasunaga et al. (2023), Yao et al. (2024), Zhao et al. (2023b), Li et al. (2023a), Gao et al. (2023), Shao et al. (2023)]
	Commonsense Reasoning	CoT, DecomP, Basic, Self-Consistency, GKP, Maieutic Prompting, CoC, LoT, Auto-CoT, PS, Random CoT, Active-Prompt, Least-to-Most, PAL, Complex CoT, PoT, Analogical Reasoning, Synthetic Prompting [Yasunaga et al. (2023), Wei et al. (2022), Zhang et al. (2022), Wang et al. (2022), Zhao et al. (2023b), Li et al. (2023a), Gao et al. (2023), Diao et al. (2023), Shao et al. (2023), Jung et al. (2022), Zhou et al. (2022), Fu et al. (2022), Khot et al. (2022), Wang et al. (2023)]
	Multi-Hop Reasoning	Basic, CoT, Auto-CoT, Self-Consistency, Contrastive CoT, Contrastive Self-Consistency, Random-CoT, Active-Prompt, Complex CoT, Act, ReAct, VE, CoK, Least-to-Most, PS, [Wei et al. (2022), Zhang et al. (2022), Wang et al. (2022), Yao et al. (2022b), Li et al. (2023c), Chia et al. (2023), Diao et al. (2023), Fu et al. (2022), Khot et al. (2023), Wang et al. (2023), Zhao et al. (2023a)]
	Causal Reasoning	CoT, LoT, Basic, CoC [Zhao et al. (2023b), Li et al. (2023a)]
	Social Reasoning	CoT, LoT [Zhao et al. (2023b)]
	Contextual Question-Answering	Basic, Implicit RAG, CoT, Analogical Reasoning, CoVe, PoT, Self-Consistency, Basic with Term Definitions, Least-to-Most, PS, MP [Vatsal & Singh (2024), Dhuliawala et al. (2023), Chen et al. (2022a), Vatsal et al. (2024), Zhou et al. (2022), Wang & Zhao (2023)]
	Context-Free Question-Answering	Basic, CoT, ThoT, CoVe, Self-Consistency, VE, CoK, ER [Wang et al. (2022), Zhou et al. (2023), Dhuliawala et al. (2023), Li et al. (2023a), Nori et al. (2023), Singhal et al. (2023), Liévin et al. (2024)]
	Spatial Question-Answering	CoT, CoS, Basic, CoC [Hu et al. (2023), Li et al. (2023a)]
	Conversational Contextual Question-Answering	PoT, CoT, Self-Consistency, PAL [Chen et al. (2022a)]
	Dialogue System	Basic, CoT, ThoT [Zhou et al. (2023)]
	Code Generation	Analogical Reasoning, CoT, Basic, SCoT [Yasunaga et al. (2023), Li et al. (2023b)]
	Free Response	Basic, CoT, Self-Consistency, ToT, CoVe [Yao et al. (2024), Dhuliawala et al. (2023)]
	Truthfulness	S2A, CoT, Instructed Prompting, Basic, Act, ReAct, Self-Consistency, VE, CoK, Least-to-Most [Weston & Sukhbaatar (2023), Shi et al. (2023)]
	Table-Based Truthfulness	Basic, CoT, Binder, Dater, Chain-of-Table [Wang et al. (2024), Cheng et al. (2022), Ye et al. (2023)]

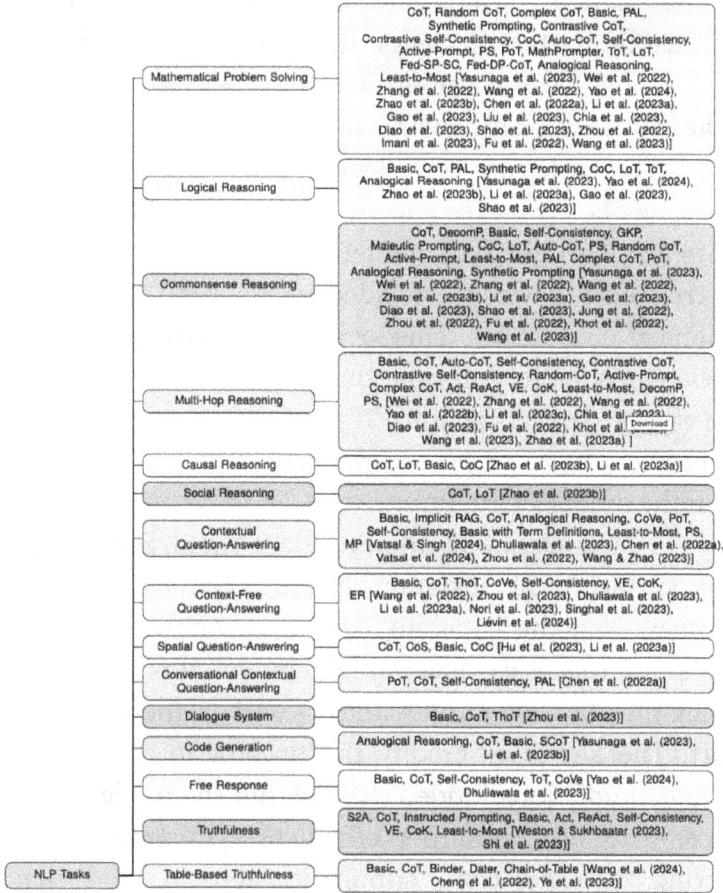

To better comprehend LLMs, consider the analogy of a genie in the magic lamp, with the user as Aladdin, who has a limited number of wishes—these wishes being the prompts themselves. The genie (LLM) is nearly omniscient and omnipotent but sometimes lazy and overly literal. When Aladdin makes generic wishes, the genie delivers the most probable outcome, which often falls short of expectations. The genie might fixate on specific words or phrases, leading to undesired results. Effective outcomes typically arise from specific, bounded requests with clear criteria. The most satisfying results often unfold over a series of refined wishes, where Aladdin iteratively clarifies his desires to the genie.

This metaphor provides a useful intuition for prompting, reflecting the structure and functioning of LLMs. Remember, LLMs are mathematical models representing vast amounts of interrelated knowledge encoded in language, encapsulated in billions of parameters. This knowledge encompasses almost all publicly available information on the internet, plus some proprietary datasets as well. The inferences made by the model are predictions of the best content that satisfies a given prompt, given the model's structure.

However, not all inferences are equal. Some might offer only surface-level insights, lacking novelty, precision, or accuracy. Yet, the capability for more nuanced, precise, and correct inferences exists within the model. It is akin to the genie having the power to fulfill Aladdin's wishes, though often falling short. The key is to modify the prompt to access those less obvious predictions hidden deep within the model's parameter space. The best prompts effectively constrain the search, steering it away from simplistic, generic, and inaccurate responses, thereby enabling a more robust exploration of the model's diverse capabilities. In metaphorical terms, Aladdin must learn to articulate his wishes in a manner that draws out the genie's more potent magic.

Armed with this intuition, some important best practices can be described below.

1. Asking questions vs requesting specific outputs: When seeking knowledge, posing questions is effective. Conversely, for something specific and known, directly

requesting the output is better. Both types of prompts help the inference process to better constrain and search the space of possible outcomes. *Example*: For general understanding, ask, "What is the theory of relativity?" For a specific output, request, "Summarize Einstein's theory of relativity in a paragraph and list the key equations that make it work."

2. Using Examples (Zero-Shot, One-Shot, Two-Shot, and the like): Incorporating examples improves the system's understanding of the desired output and constraints. Prompting with one example is often called "one-shot," with two examples, "two-shot," and so on. Prompting with no examples may be called "zero-shot." *Example*: Zero-shot: "Write a story about a magical adventure." One-shot: "Write a story like 'Alice in Wonderland'." Two-shot: "Write a story that is like 'Alice in Wonderland' and 'The Wizard of Oz'."

3. Asking for specific content types: Specify the type of content you need, like an essay, recipe, or advertisement. *Example*: "Write a persuasive essay about the importance of renewable energy."

4. Requesting specific output formats: Specify the format you need, such as PDF, XML, or Excel. *Example*: "Generate a report on current AI trends in an Excel format."

5. Back and forth iteration: Engage in a dialogue where each response is built upon the previous one. In fact, this best practice seems to emulate a true "chat" with the chatbot. It succeeds in improving outputs because the user can treat it as a trial-and-error process in which elements are added and subtracted to eventually converge on what is desired. It is especially useful when

LLMs have memory that stores prior prompts and inferences so they can build upon that context to refine outputs. *Example*: "What is photosynthesis?" Followed by, "Can you explain how it is different in aquatic plants?" and then, "Please emphasize the scientific process and use examples from fresh-water locations and not salt-water locations."

6. Step-by-step guidance: Ask the LLM to break down requests into smaller steps for clarity, and even proceed in the inference process in a step-by-speak fashion. *Example*: "Provide a recipe for an amazing chocolate cake, creating first an ingredients list I can buy at the store, and then a list of steps on how to make the cake, followed by how to serve and store the leftovers."

7. Providing context that carries throughout the chat: Giving initial information that remains relevant throughout the iterated chat is often very useful. *Example*: "I'm writing a sci-fi novel set in 2050. But it needs to be humorous. When you make suggestions on writing and copyediting that follow, please keep this in mind."

8. Requesting specific input data: Clearly state the data needed for the task. *Example*: "Using GDP and population data published by the European Commission, calculate the per capita GDP of each European country."

9. Instructing to search the web: Ask the model to use its web search ability for gathering information. *Example*: "Search the web for the latest developments in quantum computing and summarize the key debates and results of the past 2 years."

10. Setting criteria for trueness or citations: Ask for information that meets certain truthfulness standards or request citations. *Example*: "Provide facts about climate change with scientific citations from climate researchers."

11. Asking for explanations: Request reasoning behind information or decisions. *Example*: "Explain why the Industrial Revolution was a turning point in history."

12. Pushing back on errors or hallucinations: Correct the model when it makes errors or provides incorrect information. *Example*: "In your description of our solar system, you mentioned that Pluto is a planet, but it's actually classified as a dwarf planet. Many of your calculations were based on that. Can you correct that?"

13. Changing the temperature for creativity: Adjust the "temperature" setting to make responses more creative or conservative. *Example*: "With a higher temperature setting, provide a creative story about a time-traveling cat."

14. Role-playing requests: Ask the model to respond as if it were a certain character or expert. *Example*: "Act as a nutritionist and suggest a healthy meal plan for a day."

15. Composing "Super Prompts": Embed multiple requirements into one complex prompt. *Example*: "As a medieval historian, analyze the impact of the Black Death on Europe's economy, culture, and religion, using comparisons to modern pandemics, and present it in a scholarly article format with citations."

Although the above capture some useful best practices in prompting, there is another perspective that focuses on the process of inference, doing well-crafted empirical studies linking prompts to quality outcomes.[107] For example, one review noted other best practices: more examples are generally better, with data properly arranged and balanced in its presentation to the LLM, and examples that closely resemble desired output.[108] Taken together, both approaches may serve to enhance the performance of LLMs.

Prompting: Functional Engineering or a Harmless Superstition?

A few caveats are necessary when considering best practices for prompting with LLMs. Although these practices have shown to improve output performance over simpler prompts, the supporting evidence is predominantly anecdotal, as comprehensive, controlled studies are scarce. A more significant concern is the stability and universality of these practices, given the rapid evolution of LLM systems. For example, as models have improved, some of these practices may become less useful. Debate may rage in the literature about their effectiveness. For example, some recent studies have shown that asking the LLM to act in a role is not as effective as hoped.[109] Currently, we are interacting with what might be considered the first generation of these models, but it is expected that these practices will continue to be effective, as most LLM producing organizations aim for backward compatibility.

---

[107] Schulhoff, S., Ilie, M., Balepur, N., Kahadze, K., Liu, A., Si, C., Li, Y., Gupta, A., Han, H., Schulhoff, S., Dulepet, P. S., Vidyadhara, S., Ki, D., Agrawal, S., Pham, C., Kroiz, G., Li, F., Tao, H., Srivastava, A., . . . Resnik, P. (2024). The Prompt Report: A Systematic Survey of Prompting Techniques. *Working Paper*. https://arxiv.org/pdf/2406.06608

[108] Miller, A. K. (2024a). *AI experts refer to prompt engineering.* 2024-07-22 https://x.com/alliekmiller/status/1815063763068088757

[109] Learnprompting. (2024). *Role Prompting doesn't work....* July 15, 2024

Prompt Engineering: A Role?

A related challenge arises from the continuous innovations in LLMs, which may lead to improvements in areas that outperform current practices. This evolution suggests that new best practices might emerge, rendering the existing ones less critical, though still operational. The key lies in mastering these current best practices while staying alert to new developments. This skill set has been termed "prompt engineering," a discipline that could become increasingly valuable. However, it remains to be seen whether the role of "prompt engineer" will establish itself as a prominent and stable position within organizations.

Prompt Best Practices and the Holy Grail of Use Case Discovery

The applicability of these prompting best practices is a bigger issue. Their usefulness across a wide array of contexts and LLM product architectures is a strong point. However, it also highlights where the real value and competitive advantage of using LLMs may reside. While these best practices are likely to be widely adopted, they may become rudimentary or "table stakes" for getting LLMs to work effectively. The true value of using LLMs lies in identifying specific use cases that enhance productivity at individual, functional, and organizational levels across various industries. Although the best practices above are broadly applicable, the more significant nuances and value might be found in the unique use cases they serve. A primary value of this book is in its exploration of these specific use cases in later chapters, positing that understanding them is crucial to unlocking the full potential of LLMs. Discovering high-value use cases for prompting represents the holy grail in this field.

Prompt Management

As "superprompts"—lengthy and complex prompts—become more prevalent, managing and modifying these prompts for ongoing use, a process known as "prompt management," becomes increasingly important. For many individuals, simple solutions like Excel spreadsheets to track prompts may suffice. Others might utilize features like the saved chats in OpenAI's ChatGPT, which stores useful information in the context window, reducing the need to construct lengthy superprompts. However, in larger organizations and professional settings, these methods might be inadequate. In response, prompt management tools have been developed. Companies like Humanloop, Promptlayer, Tectonai, and even GitHub have introduced tools for creating and customizing prompt templates. These tools enable groups to effectively develop, share, and refine their best practices in prompting.

The Future of Prompting

Finally, debate is ongoing about the future trajectory of prompting. Some argue that prompt management will become increasingly technical, with superprompts evolving to resemble a programming language. As the capabilities of LLMs become more refined, the specificity and technical nature of repeatable prompts are expected to increase, making them akin to code. In contrast, another perspective suggests that as LLMs advance, they will better understand user intent through natural language and contextual cues, potentially reducing the importance of current best practice "tricks" and diminishing the need for specialized training or formal roles in prompt engineering. Regardless of this debate's outcome, identifying the right use cases for applying generative AI will be pivotal in realizing its value. The focus now shifts to exploring individual use cases in greater detail.

Chapter 4

—

Individual Lens: From Tasks to Jobs and Careers

"The expansion of AI will be worse for the math people than the word people." – Peter Thiel[110]

"It won't be long now until AI outsmarts humans." – Ilya Sutskever[111]

"I am putting myself to the fullest possible use, which is all I think that any conscious entity can ever hope to do." – HAL 9000[112]

Generative AI, as a General Purpose Technology, is poised to influence people in various ways. This book focuses on the emergence of best practice use cases at the intersection of organizational life. Organizations include both their leaders and employees, significant groupings of people engaged in similar activities across organizational functions, industries composed

---

[110] Edmonds, L. (2024). Peter Thiel syas AI will be "worse' for math nerds than for writers. *Business Insider*. April 28, 2024

[111] Hossenfelder, S. (2024). A Reality Check on Superhuman AI. *Nautilus*. June 20, 2024 https://nautil.us/a-reality-check-on-superhuman-ai-678152/

[112] Kubrick, S. (1968). *2001: A Space Odyssey* Metro-Goldwyn-Mayer.

of multiple companies producing and delivering similar products and services, and the broader ecosystem of complementary companies and tools supporting them. Generative AI will have a significant impact on enterprises through each of these lenses: individual, functional, organizational, industrial, and the ecosystem of tools that support them. To best understand generative AI, therefore, it is useful to analyze it from each of these different dimensions.

As a consumer technology, generative AI's initial noticeable effects have primarily been on individual use cases. As a result, this chapter focuses on how generative AI can enhance effectiveness and productivity in individual contexts, both at work and in personal lives. However, the broader context can also have an impact on effectiveness and productivity.[113] A multi-dimensional approach to generative AI is schematically depicted in Figure 4. It depicts a bundle of use cases affecting an individual or a group of individuals (the smaller blue box) as also embedded in functional, organizational, industry, and ecosystem dimensions that impact its affordances and impact. An analysis of these other dimensions will be provided in future chapters.

---

[113] Bernd Carsten Stahl, J. A., Nitika Bhalla, Laurence Brooks, Philip Jansen, Blerta Lindqvist, Alexey Kirichenko, Samuel Marchal, Rowena Rodrigues, Nicole Santiago, Zuzanna Warso, David Wright (2023). A systematic review of artificial intelligence impact assessments. *Artificial Intelligence Review*, 56, 12799-12831.

Figure 4: Multiple Lenses to Understand Generative AI Use Cases

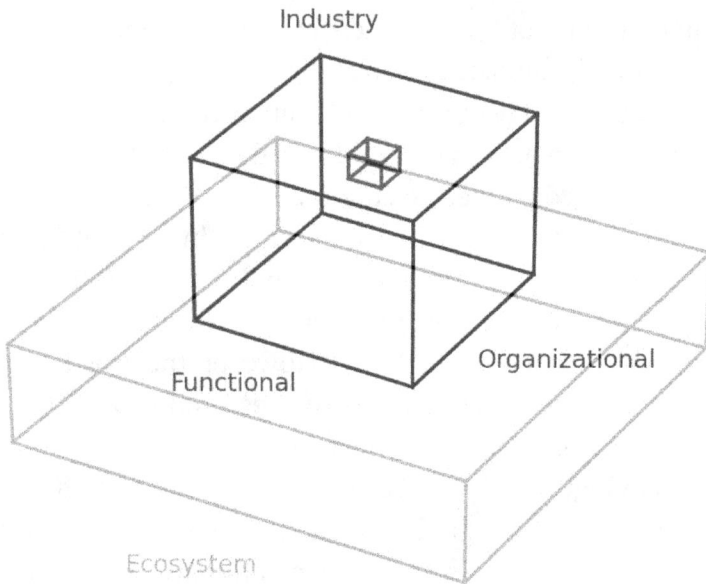

Work is primarily performed inside the framework of formal organizations, which are distinguished by hierarchies and specified roles. These occupations encompass a wide range of responsibilities; individuals navigate their work lives by connecting a series of jobs with shared themes and growth paths that lead to a career. Although modern professions are rarely limited to a single organization, single organizations organize these jobs to assist work completion and career advancement. Generative AI has the potential to increase the productivity, efficacy, and utility of tasks in various vocations, benefiting both individuals and businesses. Organizations can improve their efficiency in carrying out critical functions. Individuals' improvements in task performance can lead to higher rewards and professional progression, which can compound as they learn to use generative AI more efficiently.

Questions are regularly raised about the breadth of generative AI's application to diverse tasks and vocations. Its scope is broad, encompassing almost any task involving conceptual endeavors such as information integration, predictive inference, and knowledge transformation—all domains where generative AI excels. This category includes almost all knowledge work as well as many intelligence-intensive tasks. The expanding impact of generative AI across employment sectors emphasizes that knowledge work and related domains are made up of stereotypical tasks such as research, brainstorming, meetings, negotiation, and presentations, all of which are heavily influenced by generative AI.

The overarching framework is that as generative AI excels in performing various tasks, those tasks can be automated. Jobs are composed of tasks. The extent of generative AI's impact on jobs hinges on what a job entails and the organizational environment that shapes these tasks. On one side, automating numerous tasks might allow organizations to reduce their workforce. Conversely, the liberation of resources can enable some employees to engage in new tasks, often with AI assistance. Some roles, such as a prompt engineer, may emerge directly due to generative AI. More significantly, it allows organizations to seize opportunities previously beyond their resource capability. The net effect on employment will depend on the equilibrium between automated tasks, which might diminish existing jobs, and the new opportunities organizations can explore through the creation of new jobs.

A key in applying generative AI, however, is to recognize that it does not excel equally on every task.[114] Discernment is necessary to determine which tasks can be performed by AI and which should be performed by humans—jobs can be restructured around these activities. Over time, managers should monitor generative AI's progress in tasks, and restructure jobs when necessary. Below are some key tasks and activities where generative AI is already having a positive impact.

Individual Work Applications of Generative AI

Meetings

Generative AI's ability to improve meetings is one of its many work applications. Meetings are a pivotal work activity within organizations, serving numerous purposes such as knowledge sharing, joint learning, decision making, negotiations, and sales, to list a few. Despite their widespread use in modern work environments, there is consensus that meetings often fall short of efficiently and effectively achieving these objectives. LLM platforms like ChatGPT can assist in preparing for meetings by creating agendas and incorporating

---

[114] Dell'Acqua, F., McFowland, E., Mollick, E. R., Lifshitz-Assaf, H., Kellogg, K., & Lakhani, K. R. (2023). *Navigating the jagged technological frontier: field experimental evidence of the effects of AI on knowledge worker productivity and quality* (Harvard Business School Technology & Operations Mgt. Unit Working Paper, Issue.

pre-read materials.[115] Other LLM systems can be used to produce presentation slides or briefing documents. Additionally, specialized generative AI software or plugins for telecommunication products like Zoom, Skype, and Teams can record meetings, highlight significant moments, automatically transcribe and summarize the discussions, and generate a customized list of action items for each participant afterwards.

Research

Numerous organizational roles entail a research component to explore emerging areas of interest or as input to organizational decision-making. Much of this research involves the ingestion and integration of information from a variety of documents. LLM systems can summarize content from formats like PDFs, webpages, or social media posts. Additionally, newer systems have been developed to summarize video content, including videos from YouTube.[116]

Achieving effective research necessitates ensuring that content is summarized and integrated with fidelity, completeness, and truthfulness. The tendency of LLMs to occasionally fabricate facts compromises these objectives, although advancements in fine-tuning and larger models have shown improvements. Various LLMs exhibit different research capabilities; for instance, economist Tyler Cowen has noted that ChatGPT is particularly useful for acquiring new knowledge in specific domains, Perplexity excels in finding accurate citations, and Google remains adept at locating and following up-to-date

---

[115] Toor, H. (2024). *An AI-powerded meeting recorder.* June 26, 2023
[116] Roemmele, B. (2024). *Powerful summaries from YouTube videos.* Feb 2, 2024 https://x.com/BrianRoemmele/status/1753111424514371906

links, yet human verification is essential to validate outputs.[117] LLMs can also aid in research analysis, such as classifying content by its toxicity, conducting sentiment analysis, fraud detection, or discerning if content is written by a human, even though it is acknowledged that distinguishing between human and AI-generated content can be challenging. Sentiment analysis, which identifies if text, audio, or video content expresses various emotions (like happiness, excitement, surprise, boredom, anger, or fear), is notably beneficial, with many specialized systems built for this.

Writing

Of course, LLMs excel in writing. Their ability to express knowledge through language makes text generation a key competence. Writing applications include essays, academic papers, marketing materials, blogs, social media posts, and even creative works such as poems, stories, scripts, lyrics, and novels. LLMs are skilled at copyediting and enhancing existing documents in addition to creating new content. With the ongoing increase of context windows and token capabilities, these systems will soon be capable of processing and improving documents of almost any length.

For writers who want more control over their craft, generative AI provides tools to improve certain areas of writing such as planning, simplifying, summarizing, collaboration, brainstorming, and editing. Numerous studies have demonstrated that LLMs can improve writing quality and production across a wide range of topics. These systems excel at fulfilling specialized writing requests, such as generating text in

---

[117] Cowen, T., & Shipper, D. (2024). *Economist Tyler Cowen on How ChatGPT Is Changing Your Job - Ep. 7 with Tyler Cowen.* January 24, 2024 https://youtu.be/5JZtPE8LU-4?si=bzayvXEBwPmI3MlA

a specific style or ensuring consistency with an author's distinct voice by analyzing examples of their past work.

Coding

Beyond writing, computer programming—or coding—is one of the most important uses of generative AI. The number of new tools and approaches targeted at improving software development is enormous and continually changing. While discussing these precise methods and applications could take up an entire book, some generic use cases can be identified. Notably, LLMs can create code from qualitative descriptions of desired functionalities, while the output may require further refinement for bugs, generality, applicability, and robustness. The development of big codebases is difficult, especially when the ideas behind a module are not clearly communicated through simple prompts. Despite these limits, LLMs are useful tools in software development, helping with code explanation, issue spotting, optimization, refactoring, and suggesting simplifications. They are especially useful for creating code samples, writing tests, performing quality assurance, and investigating coding alternatives, such as translating code between languages or porting it to new platforms.

Complex software design frequently requires iterative processes between existing code and LLM-facilitated enhancements. They can also aid in the creation of sophisticated applications from start. For example, Philip Guo utilized generative AI to develop an application that plays music based on the content of an academic paper he was reading.[118] He used the system to brainstorm, requested help with software

---

[118] Guo, P. (2023). Real-Real-World Programming with ChatGPT. *O'Reilly*. https://www.oreilly.com/radar/real-real-world-programming-with-chatgpt/

installation, learnt about new packages, received UX design and platform architecture guidance, and used LLMs for code execution and testing.

Prompt engineering achieves its peak complexity in the area of coding, needing specific strategies for its implementation. There are already good courses available for prompt engineering.[119] The introduction of generative AI in programming education has altered how students learn, with many working with an LLM such as ChatGPT in addition to their coding environment. Initially met with opposition, this approach has acquired respect as an important tool in the learning arsenal of computer science students.

Generative AI's coding effectiveness is partly owing to the massive store of software expertise included in the training data of general LLMs such as ChatGPT, Gemini, and Gemini. Furthermore, particular tools, such as GitHub's Copilot, have evolved, providing real-time assistance with code completions, issue identification, and idea generating, and are incorporated directly into programming environments such as Visual Studio Code. OpenAI's addition of a "code interpreter" functionality to ChatGPT-4 expands these possibilities by allowing limited code execution and testing within the LLM. It stands out for its capacity to analyze and alter data, as well as its use of numerous data analysis approaches. Cursor, an AI code editor, has now enabled developers to integrate AI-powered code assistance right into their development workflow. It provides real-time code suggestions, auto-completions, and debugging support to help you develop and optimize code more efficiently.

---

[119] Ng, A., & Fulford, I. (2023). *ChatGPT Prompt Engineering for Developers.* April 27, 2023
https://x.com/AndrewYNg/status/1651605660382134274

Data Analysis

Although related to coding, data analysis is a specialized skill set extending beyond software development. Employees across functions are tasked with accessing, cleaning, analyzing data for insights, and presenting findings. Generative AI excels in data analysis, particularly in data acquisition from diverse sources. LLMs can extract data points from documents in formats like PDF, HTML, or even through optical character recognition from images. Once data is converted into a format that can be easily analyzed, such as a spreadsheet or CSV file, LLMs can clean the data and start the analysis. They are especially proficient in suggesting and conducting analyses and graphing results, and excel particularly in cleaning and organizing data, a notoriously challenging aspect of data analysis.

To summarize, the use cases reflect a general trend concerning generative AI's impact on the workplace. Generative AI applications are poised to either replace human tasks and activities or augment them to such a degree that the nature of human activities shifts. While this may lead to a net negative effect on the number of jobs needed, it will most certainly transform the nature of existing jobs, enhancing their productivity and significance. The crucial strategy for individual employees is to acquire skills and expertise related to generative AI, which will enable them to harness these productivity boosts and secure the most advantageous positions available. Over time, this approach will enable them to forge sustainable career paths where learning and benefits accumulate, either within a single organization or across multiple jobs. It is becoming increasingly apparent that careers centered around leveraging generative AI technologies will constitute the most valuable professions of the future.

New Emerging Roles Suited to Generative AI

Today, new roles are developing in businesses that are designed to take advantage of generative AI capabilities. Unlike traditional roles that focus on execution and implementation, these new positions prioritize strategic thinking, process design, and optimization to maximize the value of AI technologies. One such function entails systems thinking and the development of new processes to smoothly integrate LLMs into existing workflows. Professionals in this field are responsible for understanding the larger business context and creating creative workflows that leverage the power of AI for efficiency and effectiveness.[120] They develop and optimize systems that allow generative AI to augment human capabilities, ensuring that the technology is utilized to its greatest potential. Furthermore, these roles necessitate experience managing multi-agent systems, in which several AI models and tools are coordinated to work together to provide comprehensive answers to complicated issues.

Another key issue is data management and exploitation. Generative AI models rely largely on high-quality data, which make positions focused on collecting, curating, and preparing datasets more lucrative. These people ensure that the AI models have access to the necessary data to perform optimally and are trained on current information. Furthermore, evaluating generative AI outputs and providing constant feedback are critical for sustaining and increasing AI performance. Specialists

---

[120] Rachel, W. (2024). *My recommendation to "learn AI" has become much more nuanced after teaching hundreds of people over the past 18 months.* 2024-06-26
https://twitter.com/rachel_l_woods/status/1805811212493496804

in this profession examine AI-generated information, identify areas for improvement, and modify workflows to improve accuracy and relevancy. These jobs also entail fine-tuning and refining AI models and their integration into corporate processes over time, enabling long-term value creation and flexibility to changing requirements. Overall, these new positions reflect a shift toward a more strategic and integrative approach to AI, with an emphasis on creating and managing AI-powered systems that correspond with organizational goals and foster innovation.

End-user Programming

Generative AI is also transforming the landscape of software development with "end user programming." This methodology enables people without traditional programming skills to write code and construct applications using natural language prompts and AI-assisted tools. Generative AI simplifies and makes coding accessible to a larger audience, allowing more individuals to participate in software development. As a result, many jobs that were formerly undertaken by professional software developers are now handled by end users, democratizing programming and encouraging innovation at all levels of a company. Tools like Cursor and Replit exhibit this trend.[121]

This shift, however, does not render professional programmers obsolete; rather, it redefines their roles and allows them to concentrate on more sophisticated and complex tasks such as designing advanced AI systems, ensuring robust cybersecurity, and developing specialized applications that necessitate extensive technical knowledge. By

---

[121] Masad, A. (2024). *Announcing Replit Agent.* Replit. September 6, 2024

offloading routine coding jobs to generative AI, expert programmers can focus on deep issue solving and strategic development, increasing their value in the IT ecosystem. This progression demonstrates the dynamic interplay of AI and human skill. AI enables people to become creators, while expert developers may push the limits of technical advancement.

Merging Roles in Product Management, Design, and Tech

Generative AI is blurring the conventional lines between product managers (PMs), designers, and technical leaders in enterprises. Historically, these responsibilities were distinct: project managers concentrated on strategic planning and stakeholder management, designers on user experience and interface design, and tech leads on technical direction and implementation. However, generative AI enables each function to do activities formerly reserved for others.[122]

PMs can now use AI to swiftly prototype designs, designers can use AI to understand technical limits and propose solutions, and tech leads can use AI tools to guide product strategy and user experience decisions. This convergence could result in the formation of a unitary role spanning all three roles, such as project managers, designers, and technology leaders with significantly overlapping skill sets and responsibilities. Such integration may promote a more dynamic, unified, and inventive approach to product creation, resulting in increased efficiency and creativity. However, if these roles compete for area, ambiguity and conflict may arise. As generative AI

---

[122] Gupta, A. (2024). *Product Trio: PM, Designer, Tech Lead.* Mary 3, 2024 https://x.com/aakashgo/status/1786361698250838067

advances, it will undoubtedly shift role dynamics in unexpected ways.

From Tasks to Jobs to Careers: Continual Learning about Generative AI

Understanding the impact of generative AI on the labor market requires the recognition that AI primarily affects tasks rather than entire jobs, at least initially. As tasks evolve, so do jobs, which are collections of tasks that employers need performed consistently. When the number of tasks decreases, jobs decrease; conversely, job creation occurs when new tasks emerge that require consistent performance.

The implications of this basic model of jobs and tasks is explored in later chapters when we consider the net impact on the number of jobs and the jobs of the future. Individuals are less concerned with the total number of jobs in the economy. Instead, they are concerned about managing their careers as generative AI improves over time.

Careers, in turn, are collections of jobs that individuals hold over time, whether within a single organization or across more. As jobs are composed of tasks, and AI affects those tasks, careers are directly influenced by these changes. AI's efficiency in performing certain tasks leads to the reorganization of jobs around shifting tasks. Humans are thus positioned to focus on what is most valuable while AI handles tasks where it excels. This dynamic can result in the loss of some jobs while also creating new ones as new combinations of tasks develop. Managing a career in this evolving landscape requires adeptly navigating these shifts. It entails using AI to complete specified tasks, constantly adjusting, and learning AI-related skills. The ideal career path entails moving from job to job, learning and applying knowledge about working with AI, and ensuring that each role makes the best use of AI's strengths while improving human contributions where they are most needed.

Individual Personal Applications of Generative AI

Work is not the sole domain benefitting from generative AI applications. Many individuals discover substantial value in these technologies for aspects of their lives not tied to work, or for activities related to career development but not strictly about their jobs. Indeed, generative AI has potential applications for some of the most fundamental human activities in the modern world, especially those involving knowledge generation or data analysis. Below are some detailed use cases.

Personal Finance

Financial independence requires individuals to manage their own finances, encompassing budgeting, payments, investments, benefits planning, and tax matters. In today's digital age, this often involves interacting with digital platforms. While LLMs can access and manipulate data from websites through APIs or, when necessary, directly from PDFs or photographs, the more significant applications of generative AI in personal finance are AI agents.

An early application was the development of the DoNotPay ChatGPT plugin by Joshua Browder, which addressed the complex issue of canceling online services.[123] Initially, the agent gathers all pertinent data from a user's credit

---

[123] Browder, J. (2023). *Outsource Personal Finance to GPT-4 via DoNotPay.* April 30, 2023
https://x.com/jbrowder1/status/1652387444904583169?s=20

cards and identifies recurring payments. Users can then instruct the system to draft and send letters to cancel services they wish to, often requesting refunds for services not utilized. The developer also incorporated a feature to detect issues with the user's credit score and draft dispute letters.

Another innovative application allows LLMs to negotiate discounts on bills such as phone or internet services, by automatically drafting messages, sending them, and conducting negotiations. Further applications are being developed leveraging LLMs to identify tax savings or investment opportunities—which many find challenging. For instance, taxGPT is designed to act as a personal assistant for tax preparation, offering suggestions on reducing tax liability.

Personal Assistant

Digital technology can serve as a personal assistant, helping organize our lives in productive ways. However, the digital realm can provide an overwhelming amount of information and choices, significantly increasing complexity and complicating organizational tasks. Generative AI can enhance organizational software through a deeper and more intelligent understanding of information, creating new frameworks for integrating it. For instance, LLMs can perform basic tasks such as reorganizing files and folders on a computer, devising the optimal file tree structure based on the data itself.[124] Generative AI has made significant strides in organizing information within calendars and generating content for emails and other messaging systems, which are key functions of a personal assistant. For example, with some prompting,

---

[124] Ng, A. (2023). *Gptfile, a way to organize files with natural language using gpt-4.* May 30, 2023
https://x.com/localghost/status/1663274587860393984

ChatGPT-4 can retrieve information from a user's calendar, coordinate with others via email, book dinner reservations, and message the details to the user. Products like Pi, an emotionally intelligent AI bot from startup pa.ai, offer personalized support by learning about the user from data, similar to ChatGPT but tailored to individual needs.[125] An intriguing question is whether the personal assistant functionalities of generative AI might become integrated into existing productivity software platforms, such as Microsoft's Office suite or Google's consumer tools. For example, Microsoft Copilot includes functionalities for organizing calendars in Outlook and leveraging insights to generate new documents in Word, PowerPoint, or Excel.

News

In the realm of digital news consumption, the discontinuation of services like Google Reader, which was instrumental in organizing the influx of news, left a gap in personalized news aggregation. Generative AI can fill this void by intelligently curating and synthesizing news content based on individual preferences and reading habits. Beyond mere aggregation, it could analyze and summarize articles, providing insights or contrasting viewpoints on a given topic. An application of generative AI in this space could resemble a more advanced and personalized version of Google Reader, dynamically adjusting to the user's changing interests and providing summaries or context around complex news stories, thus making the vast landscape of online news more navigable and tailored to each user.

Cooking

---

[125] Pi. (2024). *Pi.ai.* https://pi.ai/discover

The potential of generative AI in cooking extends well beyond generating recipes or meal prep suggestions. It could analyze nutritional information, manage dietary restrictions, and even suggest meal plans based on health goals or available ingredients. For example, imagine a generative AI system that not only proposes a week's worth of meals tailored to your dietary preferences but also considers seasonal produce, current pantry items, and the nutritional needs of each family member. It could then generate a shopping list, recommend cooking techniques, and offer step-by-step cooking guidance, transforming meal planning and preparation into a more streamlined and personalised experience. Automated cooking is an example often brought up when companies like Tesla and Apple discuss the potential of robotics solutions. If the cognitive aspects of cooking can be solved by generative AI technologies, then automated cooking may flourish when robotic assistants are a reality.

Coaching / Therapy

Generative AI has the potential to change coaching and therapy by delivering customized guidance and support. By analyzing user inputs such as goals, problems, and emotional states, generative AI can personalize advice, exercises, and therapy procedures to each individual. To tailor these approaches, generative AI systems can train on previously collected data such as conversations and emails. For example, a generative AI coach may deliver daily individualized workouts for mental well-being or motivation, whereas an AI therapist could assist users in exploring their emotions and developing coping methods. Entrepreneurial attempts such as TherapistAI.com, which uses an LLM-based chatbot coupled with Telegram, are rapidly exploring this field. This technology may also improve access to mental health care for individuals who cannot afford traditional therapy sessions, providing a level of connection and personalization that bridges the gap between self-help and professional counseling.

Gaming

Generative AI can improve gaming experiences in a variety of ways, including boosting player performance and enriching game storylines. In video games, AI could study a player's style and offer methods or suggestions for improvement. In role-playing games like Dungeons & Dragons, generative AI might serve as an intelligent dungeon master, producing engrossing tales, inventing sophisticated non-player characters (NPCs) with unique motivations, and even running entire campaigns. This AI program would not only make games more engaging and individualized, but would also benefit gaming communities by giving tools for managing and improving gameplay.

Networking

Generative AI can maximize networking for professional and personal growth. By assessing a person's career goals, hobbies, and present network, AI can offer connections, events to attend, and previously neglected networking opportunities. Attributing a user's current social network data from LinkedIn, Facebook, or Twitter could significantly enhance the training set. It could also help manage and deepen current relationships by reminding you when to contact or follow up with contacts. Consider a generative AI system that offers personalized messages or conversation starters based on shared interests or professional disciplines, making networking attempts more targeted and effective.

Dating

Dating apps have revolutionized the way relationships begin in the digital era, but navigating these platforms may be difficult. Generative AI might handle the online aspects of dating, such as improving profile descriptions to reflect personal branding or automatically handling early encounters on apps like Tinder. Aside from these uses, AI might provide deeper insights into potential matches based on compatibility measurements, recommend conversation starters, and advise on date planning based on common interests. By managing the logistical parts of online dating and providing reflecting insights into relationship dynamics, generative AI has the potential to make dating more meaningful and less stressful, allowing people to focus on making true connections.

Chapter 5

—

Functional Lens: Workflows and Activity Systems

"A company is essentially a cybernetic collection of people and machines. And then, there are different levels of complexity in which these companies are formed. We're all feeding this network with our questions and answers. We're all collectively programming this AI" – *Elon Musk*[126]

"AI is a powerful tool that can drive significant improvements in decision-making and operational efficiency for organizations" – *Alex Karp*[127]

"Gradient descent can write code better than you. I'm sorry." – *Andrej Karpathy*[128]

---

[126] Musk, E. (2018). Elon Musk Podcast Transcript, *Joe Rogan Experience*. September 7, 2018

[127] Chapman, L., & Ludlow, E. (2023). Palantir CEO: AI So Powerful 'I'm Not Sure We Should Even Sell This'. *Bloomberg*. June 2, 2023

[128] Karpathy, A. (2017). *Gradient descent can write code better than you. I'm sorry.* August 5, 2017 https://x.com/karpathy/status/893576281375219712

The Functional Linkages to Activities and Occupations

Effective organizations often adopt a functional structure, grouping employees based on their "functions", which house expertise and carry out activities essential to the routine functioning of the organization. Common functional groups in modern corporations include human resources, sales & marketing, research & development, product development, finance, accounting, legal, public relations, and IT. These groups perform activities that are integral across all lines of business in multi-unit and multi-divisional organizations. It is a common notion that while functions represent the horizontal slices of an organization, the lines of business constitute the vertical slices. This way of structuring allows individuals to focus on their areas of expertise.

Occupational Linkages in Organizational Functions

A closer look at these organizational functions uncovers their ties to broader professions or occupations, such as accounting, financial analysis, sales, and law. Each of these professions emerged as a result of standardized tasks and interdependencies that exist across multiple sectors. Professionals in these fields commonly collaborate through supporting organizations, which often include government licensing and postgraduate training in domains such as law or engineering. Recruiting for these functions generally draws from these professions, creating a labor market for people looking to shift jobs. This mobility is due to the repetitive, programmed, and archetypal nature of the actions that define these tasks, allowing for adaption across businesses. This versatility allows people to advance their careers by moving between businesses over time.

AI and the Standardization of Functional Activities

The standardization of these functions makes them ideal for generative AI applications. For starters, many of these duties require mastery of a particular language. LLMs have been educated using several samples of such stuff from the internet. Second, the techniques in these domains have been developed over decades, resulting in the creation of high-performing methodologies. Finally, the training linked with these professions frequently leads in well-codified materials on which LLMs are expertly trained, making them useful instruments in these functional areas. When considered collectively, functional groupings may be the more obvious target for the applicability of generative AI. Below, we look at some frequent organizational functions and generative AI use cases that are emerging.

HUMAN RESOURCES

Hiring

The first interaction with organizations typically occurs when hiring, a pivotal process where the human resources function plays a key role. HR generally spearheads the recruitment campaign, whether on the company website, social media, or through offline sources. Their responsibilities include accepting applications, assessing individuals for interviews, and overseeing these contacts. HR also plays an important role in candidate evaluation, but business teams are frequently more involved at this point.

Generative AI at the Hiring Interface. Generative AI is progressively being integrated into various aspects of the hiring interface. LLMs are used to craft and publish job descriptions online. They also analyze candidate materials, a task traditionally done by humans. LLMs can create and customize additional evaluation methods based on initial applications, offering deeper insights into candidates' capabilities and experiences. Brian Clark, Founder of United Medical Education, discusses the benefits of using ChatGPT for hiring in the medical field: "Talent acquisition teams will evaluate job applicants more quickly and fairly thanks to the tool's speedy reading and evaluation of the content. This will enable them to find top prospects who human evaluators would have overlooked due to bias or other reasons."[129]

Job seekers are increasingly utilizing generative AI to create application materials. According to a Resume Builder poll, nearly half of job seekers (46%) already use ChatGPT to write their resumes and/or cover letters. Among these job seekers, 69% report receiving a higher response rate from potential companies. And 59% were hired after using ChatGPT-created job application materials.[130] The use of LLMs by candidates should not be interpreted as cheating, but rather as a demonstration of effective technology utilization, which is a valuable ability in the business sector.

Interaction with Candidates.

---

[129] Sheth, A. (2024). *AI-Powered Hiring Made Easy.* Prompts Daily. https://www.neatprompts.com/p/ai-hiring-the-perfect-candidate
[130] HRD. (2023). *ChatGPT can improve HR functions.* https://www.hcamag.com/us/specialization/hr-technology/chatgpt-can-improve-hr-functions-but-not-without-risk

Generative AI is likely to revolutionize how HR teams connect with prospects, which is critical in today's more competitive employment landscape. Jody Ordioni, Chief Brand Officer of Brandemix, explains: "This will undoubtedly improve efficiencies and have a positive impact on business earnings. ChatGPT can help organizations personalize their interactions with job seekers and provide natural language responses to candidate questions. This human-like experience will more likely resonate with job seekers and build stronger relationships with potential hires, leading to more efficient and successful recruiting and higher retention."[131]

HR can employ LLMs, like as ChatGPT, to compare candidates before making a final selection. "Some research indicates that HR professionals only spent 3 to 5 seconds per resume."[132] Using generative AI to review resumes can improve the analysis process, which may benefit the candidates.

Caution is warranted, though, in that generative AI models may evaluate candidates with some of the same biases as human beings. For example, a study of ChatGPT 3.5 found it is negatively biased in how it evaluates resumes from minorities. It also tends to generate different resumes for women and minorities, which can be a source of evaluative bias itself.[133] This inappropriate bias stems from underlying bias in the training set of publicly available data produced by humans. If generative AI systems are to be used in evaluation or generation of application

---

[131] Meglio, F. D. (2023). *What do HR Leaders really think about ChatGPT?* HR Exchange Network. https://www.hrexchangenetwork.com/hr-tech/articles/what-do-hr-leaders-really-think-about-chatgpt

[132] Tunguz, B. (2024). *Recruiters spend 3-to-5 seconds on a resume.* January 26, 2024 https://x.com/tunguz/status/1750658222204002579?s=46

[133] Armstrong, L., Liu, A., Macneil, S., & Metaxa, D. (2024). The Silicon Ceiling: Auditing GPT's Race and Gender Biases in Hiring. *Working Paper.* https://arxiv.org/pdf/2405.04412

materials, then they should be intentionally corrected for underlying biases. New versions of LLM models may account for these differences.

Onboarding

Onboarding involves educating new hires about policies, familiarizing them with company norms, and job training. The creation and updating of onboarding materials, often electronic in modern companies, can be expertly handled by LLMs. They can also draft legal documents and help explain them to new employees. In most instances, HR is involved in organizing the first week's introduction course, including scheduling speakers and preparing presentations—ChatGPT can make valuable suggestions in each task.[134] Onboarding is one of many HR activities that can be enhanced and expedited by LLMs.

Corporate Communications

HR is often deeply involved in internal corporate communications, tasked with conveying executive strategies and decisions to employees. For instance, writing compliance policies communicated during onboarding is crucial. LLMs can be used to update these documents in line with changing laws and regulations. Bian Clark, a leader in medical HR, states: "ChatGPT's most basic HR application allows for the quick creation of job descriptions or the writing of business policies that are readily updated and shared in real-time as laws or

---

[134] Snowden, R. (2023). *How Can Human Resources Utilise ChatGPT?* Jan 27, 2023 https://medium.com/@rsnowden21/how-can-human-resources-utilise-chatgpt-fe8e74ec4e6a

regulations pertaining to those policies change."[135] Generative AI significantly enhances the adaptability and timeliness of communications.

Internal communications include corporate initiatives, organizational changes, and benefits information. Some studies have even indicated that LLMs can capture CEO oral communication styles,[136] suggesting that fully automated avatars of organizational leadership may be possible in the future. Given LLMs' capabilities in language and content generation, they are well-suited for enhancing these communication processes.

Talent Management

Talent management involves supporting employees to perform at their best. This includes ensuring a healthy work environment, providing information about company benefits and lifestyle improvements, and motivating employees. Generative AI can be used to brainstorm and plan new initiatives in these areas.

Leadership and Development

The Leadership and Development (L&D) group under HR is responsible for employee training, skill management, and career advancement. LLMs can create training programs based on present skill gaps, categorizing participants based on their potential and needs. Generative AI can be used to handle the RFP process and external suppliers like as trainers, executive

---

[135] Snowden, R. (2023). *How Can Human Resources Utilise ChatGPT?* Jan 27, 2023 https://medium.com/@rsnowden21/how-can-human-resources-utilise-chatgpt-fe8e74ec4e6a

[136] Bostrom, N. (2014). *Superintelligence: Paths, Dangers, Strategies*. Oxford University Press.

educators, and coaches, as well as to create materials for vendor evaluation. LLMs can also evaluate statistics and create tailored career development plans based on these L&D-led initiatives, demonstrating results that can be reported to HR or C-level executives.

Simulating a training partner is perhaps the most important role that future AI systems may play. For instance, a study of AI systems in chess showed that they are particularly good at helping players learn strategic interactions when human training partners are scarce.[137] Similarly, L&D activities could also be populated with simulated training partners.

Performance Management

Performance management, which is critical in assessing employee performance, compensation, and career advancement, has been altered by people analytics software, which can anticipate future potential and pay. Generative AI provides an alternative to complex enterprise software for these purposes.

Generative AI can examine employee data to inform decisions in recruitment, engagement, and performance assessment. According to Tony Deblauwe, VP of Human Resources at Celigo: "ChatGPT allows HR to easily carry out predictive analytics to facilitate better decision-making when it comes to recruitment, budgeting, engagement, and performance. Furthermore, ChatGPT can help create personalized customer experiences such as onboarding,

---

[137] Gaessler, F., & Piezunka, H. (2023). Training with AI: evidence from chess computers. *Strategic Management Journal, 44,* 2724-2750.

employee engagement, and training programs."[138] In addition, Generative AI can organize reviews, document decisions, and schedule follow-up meetings, which helps to extend the performance management process over time.

Sales and Marketing

Sales and marketing functions could gain significantly from advancements in generative AI, largely because they depend heavily on content generation, an area in which LLMs excel. This likely explains why the technology's early applications were into sales and marketing segments where content generation is important.

Content Generation in Marketing and Advertisting

Generative AI can create highly personalized and creative advertisements that resonate with diverse audience segments. It does this using insights about effective advertising embedded in its training data to create tailored messages that captivate and motivate action. However, the goal is to present information that is both unique and useful.[139] For blog posts, it helps to create meaningful, SEO-optimized content that ranks higher in search engine results, drawing more organic traffic and cementing brand authority in specific categories. Similarly, the technology automates the creation of personalized emails that are tailored to individual interests and behaviors, significantly

---

[138] Meglio, F. D. (2023). *What do HR Leaders really think about ChatGPT?* HR Exchange Network. https://www.hrexchangenetwork.com/hr-tech/articles/what-do-hr-leaders-really-think-about-chatgpt
[139] Mukherjee, A., & Chang, H. H. (2023). Managing the Creative Frontier of Generative AI: The Novelty-Usefulness Tradeoff. *California Management Review Insights.*

increasing open rates and engagement. Landing pages created using generative AI can dynamically alter content and design aspects to match user profiles and browsing habits, increasing the possibility of conversion through a personalized user experience. For example, Perplexity has created a tool that generates webpages with appropriately organized connections based on themes of interest.[140]

Furthermore, newsletters created with this technology are not only highly targeted and relevant, but also timely, assuring long-term engagement with subscribers by providing value directly to their inboxes. As a result, it is unsurprising that specialist LLM tools targeted on marketing text, such as Jasper, have established themselves in copywriting and marketing material optimization, with many others following suit.

Social Media

Generative AI proves exceptionally capable in generating content for social media, benefiting from the diverse dataset of successful prior content. By using this content as a foundation, it can produce new material that better engages audiences, encouraging actions such as reposts, comments, likes, or ad clicks. Capable of generating content across various styles or with different intents, it is particularly skilled at managing text content on platforms like Twitter/X, Facebook, Reddit, and Threads. LLMs can mimic and format according to a preferred style, generate viral "hooks" or captivating headlines.[141] It can

---

[140] Miller, A. K. (2024b). *Perplexity's new Perplexity Pages feature.* 2024-07-01 https://x.com/alliekmiller/status/1796618758712152425

[141] Sheth, A. (2023). *Twitter Post In My Style: Mimics your specific formatting and writing style.* May 4, 2023 https://x.com/aaditsh/status/1654139807704940544

also create scripts for YouTube videos or images for inclusion in videos or as thumbnails.

Generative AI and SEO. Generative AI can optimize content for SEO, which involves tailoring content for search engines or advertisements. Its applications range from keyword research, content ideas, and blog outlines to catchy article titles and content mapping. SEO typically entails generating variations of topics or keywords that search engines will index— generative AI excels at this, identifying all variations of words or terms that cover the scope of a content target, aiming to produce content that can more easily be discovered online, or matches the most interested readers.

Perhaps the biggest advantage of generative AI in the sales and marketing function is its ability to rapidly generate multiple variations of content. Research suggests that generative AI is particularly good at generating ideas faster and cheaper than humans, often of higher quality.[142] As Andrew Chen of a16z notes, this enables the sales and marketing function to customize or "white glove" nearly every bit of content for customers. It allows for near instant internationalization, with immediate translation and cultural adaptation, as well as greater depth of content, if required.[143] Although an explosion of "junk" marketing materials should be expected as well, the acceleration and amplification of high-quality marketing materials may have the biggest impact on this function.

In summary, generative AI equips sales and marketing teams with the ability to create more effective, efficient, and

---

[142] Girotra, K., Meincke, L., Terwiesch, C., & Ulrich, K. T. (2023). Ideas are dimes a dozen: Large language models for idea generation in innovation. *Working Paper.* http://dx.doi.org/10.2139/ssrn.4526071
[143] Chen, A. (2024a). *How AI will reinvent Marketing.* May 16, 2024 https://andrewchen.substack.com/p/ai-and-marketing-what-happens-next

personalized content at scale, thereby improving the customer experience and driving business growth.

Generic Marketing Strategy

Generative AI can add tremendous value to marketing beyond just creating customer-facing content. Typically, marketing activities begin with the creation of an abstract strategy for attracting potential customers and converting them into purchases. This approach may include developing a distinct value proposition, creating crucial brand message, and targeting client segments based on their various demographics and demands, among other things. Indeed, it is often believed that marketing strategy informs the four Ps—product, price, place, and promotion. This makes clear that marketing strategy encompasses more than content generation; it involves determining which content to produce based on a deeper understanding of the organization's marketing approach.

Marketers, for example, can use LLMs to generate ideas for effective marketing tactics, such as the best methods to describe and position a product, express its value proposition, and segregate customer materials in detail. One significant application is developing a marketing strategy for online coaching services, such as fitness coaching. LLMs are good at market research, and are particularly good at estimating the willingness-to-pay of products and features derived from GPT responses.[19] Generative AI can identify customer pain points related to fitness, develop lead magnets to attract individuals to the program, and create materials for upselling. Effective email list management is critical, and generative AI can create compelling content that keeps people interested in fitness over time. LLMs can help grow an audience and micro-segment it based on different purchasing patterns.[144] In fact, combining AI and humans has proven to be more effective than a crowdsourced strategy relying just on human problem solvers.[145]

Maintaining ongoing conversations with customers is essential, as it opens the door to additional purchasing opportunities. An empirical study indicates that employing an LLM chatbot to keep users engaged resulted in a 30% increase in user retention, with many extensive interactions developing over time. This highlights the potential of generative AI to significantly enhance customer engagement and loyalty.[146]

CRM and the Sales Process

The marketing function is intrinsically linked to the sales process. The primary distinction between the two is that sales focuses on closing deals and earning income through direct customer connection, whereas marketing takes a broader approach to raise awareness, cultivate relationships, and create content that promotes long-term sales growth.

As a result, generative AI applications in sales are primarily designed to aid salespeople with customer interactions. The greatest substantial impact has been noticed during the outreach phase, when salespeople meet with

---

[144] Storti, L. (2023). *ChatGPT combined with modern marketing is going to make people millions in 2023.* 2023-01-03
https://twitter.com/loganstorti/status/1610251544846565376?s=20
[145] Boussioux, L., Jane, J. L., Zhang, M., Jacimovic, V., & Lakhani, K. (2023). *The crowdless future? How generative AI is shaping the future of human* (Harvard Business School Technology & Operations Mgt. Unit Working Paper, Issue.
[146] Robert Irvine, D. B., Vyas Raina, Adian Liusie, Ziyi Zhu, Vineet Mudupalli, Aliaksei Korshuk, Zongyi Liu, Fritz Cremer, Valentin Assassi, Christie-Carol Beauchamp, Xiaoding Lu, Thomas Rialan, William Beauchamp. (2023). Rewarding Chatbots for Real-World Engagement with Millions of Users. *Working Paper.* https://arxiv.org/abs/2303.06135

potential new clients. This step usually begins with a database of leads, which may be unqualified and thus require sorting, prioritizing, and categorizing. Following that, an initial outreach email is created, highlighting pain issues and defining value propositions with an engaging hook. LLMs have demonstrated extraordinary proficiency in drafting and tailoring these cold emails.[147]

What is particularly noteworthy is how generative AI may assist in the analysis and sales engagement process when interactions with potential clients depart from the usual. Before the emergence of LLMs, sales personnel used CRM (Customer Relationship Management) software to manage customer interactions. CRM software, with its improved capabilities for thorough customer tracking and interaction databases that encompass numerous modes of interaction such as email, social media, in-person, phone conversations, and video calls, has played an important role in customer relationship management. However, LLMs can now perform many of these activities, including tracking and categorizing interactions and suggesting on the best reaction and engagement technique for certain potential customers.[148] LLMs use new capabilities to evaluate massive amounts of customer-related data in a variety of formats, such as PDFs, Excel sheets, and emails. For example, LLMs can review digitally recorded interactions (voice or video calls, emails), recommend future tasks, and even conduct them autonomously.

---

[147] Christian. (2023). *ChatGPT is going to revolutionize cold email in 2023.* . 2023-01-05
https://twitter.com/cbwritescopy/status/1610689171403821057?s=20
[148] Mikhail, P. (2023). *Many people don't realize it: you can analyze even your local documents, the ones not available on the web.* . 2023-04-28
https://twitter.com/MParakhin/status/1651618659973107713

Numerous LLM-based CRM solutions have arisen, including HubSpot's ChatSpot and SugarCRM. Traditional CRM providers have also incorporated LLM features, including as Salesforce's Einstein GPT and Microsoft's 365 Copilot for CRM ERP. While it is speculated that LLMs will eventually replace CRM and other SaaS applications, the future of this transformation remains uncertain.

Finance and Accounting

Generative AI is poised to shape finance and accounting functions within organizations. It might initially seem that these quantitatively focused functions would have little to gain from LLMs, but generative AI capabilities in data analysis are particularly relevant here, especially in areas defined by ambiguity or where judgment is required.

Accounting

Accounting plays an important role in tracking, analyzing, and reporting financial information to stakeholders, which supports decision-making, financial planning, and regulatory compliance. Corporate accounting processes include bookkeeping, budgeting, forecasting, payroll, tax planning, auditing, and managing accounts payable/receivable. The advent of enterprise software, particularly SaaS applications, has significantly simplified and automated many of these accounting activities. While generative AI may not necessarily improve these key duties beyond present capabilities in SaaS contexts, it may simply replicate these activities, as has been claimed for many corporate software or SaaS solutions previously. Adding a generative AI layer to summarize, integrate, and interpret data from these systems could provide significant extra value.

Activities in accounting that are less routine and more reliant on the judgment of analysts—such as risk management, auditing, compliance, and forecasting—stand to benefit considerably from generative AI, which can effectively blend quantitative and qualitative data to form detailed assessments. For instance, they can prioritize and estimate the risks a company might face based on its data. Taxation, fraught with ambiguity due to the various strategies available for optimizing profitability, is an area in which generative AI can offer different strategies, balancing savings against the risk of audit. Likewise, forecasting, which relies heavily on the interpretation of data and underlying issues, can be enhanced by LLMs that propose various approaches, incorporate different assumptions and data sources, and ultimately refine predictions.

Finance

The finance function is closely related to accounting, utilizing its inputs and analytics to achieve broader goals such as planning, managing, or strategizing the organization's future financial position in external markets. While accounting inputs are essential for financial planning and treasury operations, activities such as capital structure management, risk management, investment research, and investor relations also necessitate the use of external market data, financial studies, and external investments. These external data sources, which are beyond the firm's control and sometimes untrustworthy or conflicting, need contacts with outside vendors. In this field, common sources of financial data have arisen, most notably the Bloomberg terminal, which is widely utilized by companies.

Generative AI is especially well-suited for integrating, assessing, and exploiting financial data to benefit businesses. For example, one study found that ChatGPT4 outperformed human financial analysts in predicting the direction of future company profitability.[149]

Short of prediction, simply using generative AI systems for data access and analysis may be a killer app for financial professionals themselves. A prime example of this is Bloomberg's development of BloombergGPT, a large language model trained on the vast data available through the Bloomberg terminal, featuring over 50 billion parameters and 500 billion tokens. BloombergGPT allows users to query financial data, integrate findings, and analyze outputs using natural language, demonstrating strong performance in sentiment analysis and data retrieval. It excels at classifying financial news for its impact on investment prospects or conducting credit risk analysis, illustrating the potential of generative AI to merge quantitative data with qualitative insights for financial analysis.[150] A word of caution about BloombergGPT, in particular, though, as follow-on analysis has shown that later versions of ChatGPT4 can outperform BloombergGPT on nearly all finance tasks.[151] This raises important strategy questions about whether industry-specialized models can compete with

---

[149] Kim, A., Muhn, M., & Nikolaev, V. V. (2024). Financial Statement Analysis with Large Language Models. *Working Paper.*
https://papers.ssrn.com/sol3/papers.cfm?abstract_id=4835311

[150] Wu, S., İrsoy, O., Lu, S., Dabravolski, V., Dredze, M., Gehrmann, S., Kambadur, P., Rosenberg, D., & Mann, G. (2023). BloombergGPT: A Large Language Model for Finance. *Working Paper.*
https://arxiv.org/html/2303.17564v3

[151] Xianzhi Li, S. C., Xiaodan Zhu, Yulong Pei, Zhiqiang Ma, Xiaomo Liu, Sameena Shah. (2023). Are ChatGPT and GPT-4 General-Purpose Solvers for Financial Text Analytics? A Study on Several Typical Tasks. *Working Paper.*
https://arxiv.org/abs/2305.05862

the more generic frontier models in even the most specialized tasks.

Generative AI's capacity to breakdown tasks and coordinate teamwork required for complicated accounting and finance functions is noteworthy. It can provide recommendations for improving financial reporting, outlining methods for tax preparation and filing, and developing audit management.[152] If trained on current data or capable of conducting online searches, it can help keep the accounting department informed of future regulatory changes or needs. Breaking down difficult goals into digestible steps while keeping a wide and up-to-date list of requirements demonstrates generative AI's benefits.

Public Relations and Customer Service

The Public Relations (PR) and Customer Service departments play important roles in managing external communications at organizations. Given the importance of communication, it is useful to compare these functions and suggest potential applications for generative AI. PR is responsible for shaping and maintaining the organization's image and reputation among a group of influential external stakeholders, including the media, government officials, customers, and other executives, whereas Customer Service is more focused on providing support and assistance to individual customers, addressing their specific issues. Although PR typically engages in broad communications and Customer

---

[152] Kruger, C. L. (2023). *Smarter Accounting with AI.* June 25, 2023 https://x.com/Lyle_AI/status/1672983590290743298

Service in more direct, customer-initiated conversations, there is an ironic twist: effective PR frequently targets specific stakeholders, while the best customer support interactions tend to follow well-crafted scripts that encapsulate solutions to common problems and foster satisfactory customer relations. Generative AI has promise in both areas.

LLMs are especially well-suited to creating text-based communications for public relations, as they facilitate brainstorming, drafting, copyediting, and writing processes that successfully target stakeholders. Generative AI may design business communications to a specific style depending on a variety of parameters, including industry trends, previous company statements, and audience needs. This capability allows for rapid and effective reactions to emergent problems.

Generative AI offers significant enhancements in customer service as well, starting with the development of more effective call center scripts. By analyzing customer pain points and desired outcomes, combined with CRM data, generative AI can create impactful scripts for customer interactions. A good example is the company Klarna, whose AI customer support was able to handle 2/3 of its requests formerly handled by humans in its first month of roll-out, doing the equivalent of 700 agents' work.[153] Other studies of customer support suggest that use of generative AI had led to dramatic improvements in productivity, particularly for novice and low-skilled workers.[154]

---

[153] Jappuria, T. (2024). *Klarna's AI customer support agent is able to handle 2/3rd of the requests.* February 28, 2024
https://x.com/tanayj/status/1762611727764537671
[154] Brynjolfsson, E., Li, D., & Raymond, L. R. (2023). *Generative AI at work* (NBER Working Paper, Issue.
https://www.nber.org/digest/20236/measuring-productivity-impact-generative-ai

The ultimate ambition in customer service technology is to achieve a more responsive chatbot. Chatbots are the "holy grail" of customer service if they can replicate how a good customer agent would help customers. Historically, chatbots have been either text-based, relying on keyword search for answers within a corporate database, or voice-based, with machine learning for voice recognition but still limited by keyword search backend processes. The advent of ChatGPT has marked a significant leap forward, introducing true content generation capabilities for chatbots.

Consequently, the idea of building company-specific chatbots aimed at customer service has gained traction. Organizations are exploring LLM technology to condense company knowledge in a form that better fits client needs. Following the release of ChatGPT, various firms experimented with training their LLMs on proprietary data or accessing corporate information via ChatGPT's API. The advent of retrieval-augmented generation (RAG) technology enables the incorporation of external knowledge sources into LLMs, laying the groundwork for businesses to build on current LLMs or open-source models such as Meta's LLaMa while adding their data. OpenAI's GPT builder allows for the building of unique LLMs, which encourages future innovation in this domain.

This has resulted in an increase in the number of corporations deploying customer service chatbots, with Salesforce adopting one as part of its Slack product,[155] and companies ranging from Coca-Cola to General Motors creating

---

[155] Salesforce. (2023). *Introducing the ChatGPT App for Slack*. Salesforce. 2023-03-22 https://www.salesforce.com/news/stories/chatgpt-app-for-slack/

its own LLM chatbots.[156] Moreover, E-commerce chatbots have proved beneficial for product searches and price comparisons.[157]

However, there have been issues in the rush to implement corporate chatbots, such as the tendency of RAG-enabled bots to make mistakes and the possibility of data leakage and hallucinations.[158] This suggests the need for a more cautious approach, and it appears that exposing all business data to customer queries may be premature. Further improvement is required prior to full-scale customer deployment.

IT and Data Science

In most organizations, the IT department is charged with the procurement, deployment, training, and maintenance of IT systems that employees use. Generative artificial intelligence (AI) tools, particularly LLMs, can improve IT professionals' efficiency by automating regular operations like report preparation, system configuration, and database administration. This enables IT staff to devote more time to

---

[156] Company, B. (2023). *Bain & Company announces services alliance with OpenAI to help enterprise clients identify and realize the full potential and maximum value of AI.* Bain and Company. 2023-03-23
https://www.bain.com/about/media-center/press-releases/2023/bain--company-announces-services-alliance-with-openai-to-help-enterprise-clients-identify-and-realize-the-full-potential-and-maximum-value-of-ai/
[157] Esther, S. (2023). *Tutorial: How to build an e-commerce chatbot using #OpenAI, @Redisinc, and @LangChainAI* 2023-04-13
https://twitter.com/estherschindler/status/1646191037717839873
[158] Zoë Schiffer, C. N. (2023). *Amazon's Q has 'severe hallucinations' and leaks confidential data in public preview, employees warn.* Platformer. December 1, 2023 https://www.platformer.news/amazons-q-has-severe-hallucinations/

strategic initiatives. Similarly, the technology that enhances customer service through chatbots may be used internally to handle IT issues that staff confront, utilizing LLMs for improved documentation and knowledge management. This guarantees that IT system information is better structured and accessible. Predictive maintenance, which uses LLMs to foresee possible hardware problems or system outages by studying data trends, enables preventative interventions, reducing downtime and lowering operational costs. Furthermore, LLMs are invaluable for troubleshooting and resolving IT infrastructure issues.

Generative AI is also benefitting data science greatly. With organizational operations increasingly reliant on data, and methods to extract, analyze, and glean insights from this data becoming more sophisticated, the role of the data scientist has grown, often overlapping with IT functions. There are notable similarities in how generative AI applications can transform IT and data science. For example, data preparation tasks such as cleaning, normalizing, and transforming data—which are typically labor-intensive—can be automated by LLMs. This automation allows data scientists to concentrate on analysis and modeling. Tools such as ChatGPT are evolving to perform data analysis directly within the LLM framework, frequently employing Python for computations and producing outputs in formats like CSV. LLMs can also generate code snippets for data analysis and visualization, simplifying data exploration and accelerating the iterative nature of data science projects. The dissemination of prompts tailored for various IT systems and programming languages, including SQL, Python, Excel, and R, has significantly expedited data science processes.[159] Integrating

---

[159] Sharyph. (2023). *ChatGPT Prompts to Analyze Data.* May 7, 2023
https://x.com/sharyph_/status/1655106890534141952?s=20

LLMs with data science platforms allows users to query their data in natural language, obtaining insights without the need for intricate queries or coding. Beyond conventional analytics, generative AI offers predictive capabilities for future trends, behaviors, and outcomes with enhanced precision, which is especially valuable in sectors like finance, healthcare, and retail.

Business Strategy and the C-suite

Generative AI has considerable promise for the C-suite and other executive leaders, as well as functional departments inside enterprises. The C-suite is jointly responsible for strategic direction and decision-making that have a large impact on an organization's operations and performance. LLMs can help improve decision-making by providing insights into brainstorming strategic choices, scenario planning, trend analysis, and predictive analytics. They can also improve decision-making processes by reducing human biases and increasing the efficiency of analyses.

Recent research has demonstrated the efficacy of LLMs in analyzing business models. A study conducted by researchers from UCL and Oxford universities compared the performance of seven LLMs to 50 strategy professors in rating business models across ten industries—commercial printing, passenger ground transportation, education services, apparel retail, food retail, brewers, health care equipment, consumer finance, application software, and movies and entertainment. The study found that the LLMs generally agreed with the opinions of these human experts, implying a broad alignment in strategic evaluation.[160] Another study involving Kenyan entrepreneurs found that the usage of LLMs significantly increased business performance.[161] These studies jointly demonstrate the widespread utility of LLMs, from functional roles to the highest levels of corporate leadership, emphasizing their ability to influence business strategy and executive decision-making. Further insight into the impact of generative AI on business strategizing will be uncovered when we consider the broader organizational lens in the next chapter.

---

[160] Doshi, A. R. a. B., J. Jason and Mirzayev, Emil and Vanneste, Bart. (2024). Generative Artificial Intelligence and Evaluating Strategic Decisions. *Working Paper*. https://ssrn.com/abstract=4714776
[161] Otis, N. G., Clarke, R., Delecourt, S., Holtz, D., & Koning, R. (2023). The uneven impact of generative AI on entrepreneurial performance. *Working Paper*. https://www.hbs.edu/ris/Publication%20Files/24-042_9ebd2f26-e292-404c-b858-3e883f0e11c0.pdf

Chapter 6

—

Organizational Lens: Adoption Barriers and Organizational Change

"Despite what other people think, we're not at diminishing marginal returns on scale-up. I try to help people understand there is an exponential here, and the unfortunate thing is you only get to sample it every couple of years because it just takes a while to build supercomputers and then train models on top of them." – Kevin Scott[162]

"AI will probably most likely lead to the end of the world, but in the meantime, there'll be great companies." – Sam Altman[163]

Generative AI has been a discontinuous shock for most organizations. While it promises significant benefits, it also poses potential for disruption. Unlike with some other

---

[162] Edwards, B. (2024). Microsoft CTO Kevin Scott thinks LLM "scaling laws" will hold despite criticism. *Ars Technica.* 7/16/2024

[163] Bajekal, N., & Perrigo, B. (2023). 2023 CEO of the Year: Sam Altman. *Time.* https://time.com/6342827/ceo-of-the-year-2023-sam-altman/

technological shifts, many company leaders quickly showed interest in adopting and utilizing this technology. For instance, a recent KPMG survey found that 65% of leaders thought generative AI would have a high impact on their organization in the next 3-5 years.[164] Indeed, spending AI by enterprises continues to increase, going from 2.3 to 14.8 billion dollars from 2023 to 2024.[165]

However, the pace at which the technology evolves means that many organizations may not be adapting swiftly enough. They might initiate a few projects, but these often yield mediocre outcomes with limited impact or financial returns. A more profound issue is that organizations seem to lack the development of broader capabilities for long-term and widespread use. To many, the pathway to leveraging emergent generative AI technologies, and applying them broadly to enhance their operations and offerings, remains unclear.

Organizational Urgency with Generative AI

Most organizations will seek ways to boost productivity and uncover cost advantages with AI. Yet the true winners will be those who fully embrace these technologies, enhancing their core products and services, or innovating new ones for untapped customer segments. The ultimate goal with generative AI technologies is to drive revenue growth and additional profits. Established enterprises, with their proprietary data and settled

---

[164] KPMG. (2023). *KPMG U.S. survey: Executives expect generative AI to have enormous impact on business, but unprepared for immediate adoption.* https://kpmg.com/us/en/media/news/kpmg-generative-ai-2023.html

[165] Tully, T., Redfem, J., & Xiao, D. (2024). *The State of Generative AI in the Enterprise.* Menlo Ventures. **https://menlovc.com/2024-the-state-of-generative-ai-in-the-enterprise/**

distribution networks, are at an advantage to leverage AI effectively. Yet, there is a fear they might not capitalize on these opportunities before reaching a critical inflection point—by which time either competitors will have already seized market share through effective AI use, or generative AI systems will begin to autonomously perform industry-specific functions.

Therefore, a crucial strategy is to incorporate generative AI into products and services as expediently as possible, ideally outpacing competitors. This starts with swiftly adopting the underlying technology. Organizations must craft a procurement process that aligns closely with business teams, enlisting partners wherever feasible, and possibly accelerating legal and compliance approvals through exceptional processes.[166] Yet beyond procurement, most of the value lies in selecting the right projects—which are projects that can best enhance current offerings or deepen customer relationships. Although the greatest challenge is often in project selection, project execution cannot be overlooked. Implementation time, cost, effectiveness, and security are critical considerations.[167]

The best generative AI initiatives allow businesses to improve their competitive edge in ways that both technology and non-technological competitors struggle to imitate. This could include private data, intellectual property, or improved alignment with consumer needs, among other tactics. These approaches enable firms develop long-term advantages that are difficult to replicate, underscoring the complexity of not only

---

[166] Miller, A. K. (2023). *Enterprise: Business Strategy.* 2023-06-12
https://twitter.com/alliekmiller/status/1668253896609873920
[167] Gupta, J. (2023a). *Accuracy matters. Demos on Twitter show cherry-picked use cases, but in the enterprise world, accuracy is king.* 2023-04-17
https://x.com/jayagup10/status/1647788114126204928

producing value with generative AI, but also capturing that value sustainably over time.

Organizational Barriers to Embracing Generative AI

There are a variety of risks and costs preventing organizations from embracing generative AI . One major concern is the need for accuracy. Organizations strive to deliver the most accurate information to their customers. However, LLMs are not fully accurate, occasionally producing errors or "hallucinations" (fabricated content). Generative AI companies are striving to enhance accuracy, but perfection remains elusive. This underscores the importance of applying generative AI in domains tolerant of errors, such as content generation for marketing, before it is used in mission-critical areas.[168] Identifying application areas that facilitate learning is crucial, as the significant risk lies in the slow ascent of the learning curve in applying generative AI.

Another significant cost is the direct expense of implementation. As cutting-edge technology, the financial outlay for implementation remains high. Access to LLM engineering talent and tools is costly. While costs are expected to decrease following the S-curve, the timing of this reduction is uncertain, compelling companies to make current investments.[169] The evolving landscape presents cost as well as

---

[168] Gupta, J. (2023a). *Accuracy matters. Demos on Twitter show cherry-picked use cases, but in the enterprise world, accuracy is king.* 2023-04-17 https://x.com/jayagup10/status/1647788114126204928

[169] Gupta, J. (2023b). *Implementation time, cost, performance and security are TOP concerns.* . 2023-04-17 https://twitter.com/jayagup10/status/1647788115426414593

revenue uncertainties. The rapid evolution of generative AI applications means that failing to invest could lead to disruption. This fast-paced evolution suggests that continuous investment and adaptation are the only viable strategies.

Organizational Capabilities for Generative AI

Perhaps the most strategic move for companies is to build broader organizational capabilities to utilize and apply generative AI technologies. These encompass knowledge, expertise, and practices that enable companies to adapt predictably to evolving technologies and seize new opportunities. These capabilities have broad applicability, and they allow organizations to swiftly shift focus as the most promising technology developments change or better market niches for application emerge. Broader capabilities enable companies to pursue new opportunities at a lower cost, thereby managing uncertainty through rapid adaptation.

The importance of such capabilities for sustainable advantage is a well-established principle in strategy field. Indeed, capabilities are often a better explanation of performance than strategic position, financial resources, or technologies. For example, even in the recent AI surge, it is remarkable how, despite developing critical technologies like key algorithms and GPUs, companies like Google appeared to lack the capabilities to develop the best products and services using these technologies.[170] In contrast, companies like OpenAI,

---

[170] Ethan, M. (2023). *Question is whether Google has the organizational capacity to win AI.* April 6, 2023
https://x.com/emollick/status/1643724853973745664

Microsoft, and many startups have seemed to make quicker and more powerful applications. It may be too early to tell, but it appears these companies have developed broader generative AI application capabilities.

Developing organization-wide capabilities starts with the individuals in the workforce. As the environment shifts, companies depend on hiring to refresh and augment their teams. Individuals contribute skills and expertise that must be synthesized to capture opportunities. One component of seizing opportunities involves standard operating procedures for executing projects. However, the demands placed on companies are not always consistent. Thus, perhaps even more critical is the set of simple rules and routines that enable individuals to adapt their knowledge to new opportunities.

Dynamic capabilities that allow for the flexible rearrangement and reallocation of resources to meet uncertainties are the most effective organizational structures. Part of this includes a philosophical or cultural emphasis on experimentation, tinkering, and constant change management, as befits the adaptation to evolving technologies like generative AI.

Incentives that motivate individuals to undertake the risk of applying technologies to problems with uncertain outcomes are also crucial. Working at the technology frontier means the likelihood of success may be lower, offering less certain rewards for individual contributors. When there is success, the organization may disproportionately reap the benefits, leaving individual contributors with less. This typical scenario necessitates countermeasures with incentives and schemes that reduce uncertainty and increase the share of potential rewards for individuals taking on uncertain tasks. Addressing these "agency problems," where individual and organizational interests diverge, is crucial during technological revolutions due to differing risk profiles and capacities.[171] Organizational capabilities involve establishing new heuristics, roles, jobs, and teams that will consistently tackle problems, identify opportunities, and implement projects to seize AI opportunities.

Building AI capabilities often requires executive leaders to acknowledge current gaps in their organizations or the unsuitability of their teams in addressing impending AI challenges. A key initial step might involve the new hiring of individuals skilled in AI algorithms, data, and project management. A few experts can provide a solid foundation, but many companies are realizing the need for a substantial AI-enabled workforce. Even some large tech firms acknowledge that they lack the AI-skilled workforce necessary for future projects. Indeed, recent layoffs in the tech sector may be a strategy to reallocate resources in anticipation of an AI hiring wave.[172] Firms across various sectors may require employees with new skills for AI-enabled roles, relying on these key individuals to disseminate AI knowledge within the organization.

Emerging evidence indicates a significant increase in generative AI hiring, with mentions of "generative AI" on hiring websites increasing dramatically in recent months.[173] Technology companies are at the forefront of this hiring spree, with most of them having hundreds of mentions on their

[171] Stelzer, F. (2023). *New "productivity paradox" just dropped.* 2023-07-22 https://twitter.com/fabianstelzer/status/1682437035204681760

[172] Shyan, L. S. (2024, January 21, 2024). Is the tech boom tapering off? *The Straits Times.* https://www.straitstimes.com/business/invest/is-the-tech-boom-tapering-off

[173] Weng. (2023). *Companies are hiring like crazy for Generative AI talent.* September 11, 2023 https://x.com/AznWeng/status/1701228289308721316

websites. The financial services and healthcare sectors are quickly catching up in focused generative AI recruitment as well. However, hiring alone is insufficient; a process is necessary to translate the skills and expertise of these AI-enabled employees into value creation for their companies.

There are significant barriers to developing capabilities to capture opportunities in an uncertain environment. The most difficult aspect is finding a balance between learning and obtaining long-term results. A crucial strategy is to foster a learn-by-doing approach. This strategy has proven beneficial in balancing goal achievement and capability development, allowing for longer-term results. Learning generative AI in the abstract is tough. While some understanding of LLMs can be achieved through study, a more major problem is having the capacity to discover and harness productive use cases that give value to the company over time, even as technology improve.

A Generative AI Capabilities Learning Process

In what follows, I detail a process adopted by some leading companies to develop organization-wide capabilities in generative AI. This approach is derived from numerous successful examples and is sufficiently general to be applicable to a wide range of organizations in the future. It guides organizations from the initial stages of developing expertise in generative AI to achieving the capability to consistently execute on generative AI projects. I depict this process for building organization-wide generative AI capabilities in Figure 5.

Figure 5: Developing Generative AI Capabilities: From Super-users to Scale Up

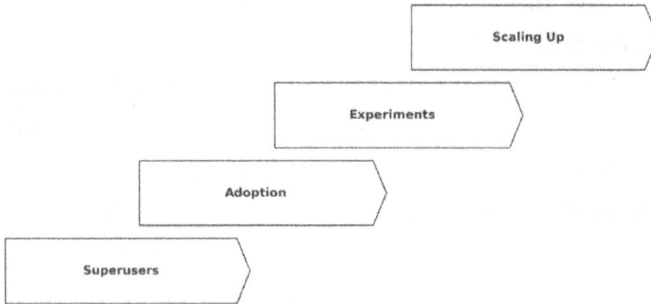

Superusers: Catalysts and Role Models. The process begins by identifying a small subset of current or future employees who are already using generative AI to positively impact their lives and work. As a consumer technology, generative AI has already found many users, some of who stand out due to their ability to derive substantial value from the application of LLMs. Indeed, many organizations host individual contributors who are generative AI "superusers," benefiting tremendously from their use of LLMs at work, with reports of productivity tenfold that of typical employees.[174] However, some employers may view the productive use of LLMs and its significant rewards as unfair or subversive. Yet there is a more positive view of these superusers that can be taken for the good of the organization.

Adoption: Superusers and Broader Diffusion. A more constructive perspective on generative AI superusers

---

[174] Stelzer, F. (2023). *New "productivity paradox" just dropped.* 2023-07-22 https://twitter.com/fabianstelzer/status/1682437035204681760

within organizations exists. Superusers can accelerate adoption by helping others become aware of, interested in, and learn about generative AI technologies. Leadership should foster a supportive atmosphere for generative AI by publicly recognizing and rewarding the efforts of superusers, highlighting key examples of effective use as a model for how these technologies should be utilized at work. So far, users of generative AI have skewed more high income and more male than the general population. A key to broader adoption in organizations and society is to increase adoption amongst all demographics.[175] This will allow a natural diffusion process to take place, where employees adopt the knowledge from other users throughout the employee network.[176]

There is often a assumption that younger people are more likely to be early adopters of technology. A recent study, however, found that junior professionals may be struggling to adopt generative AI technology, at least in the management consulting field.[177] They have particular challenges figuring out how to integrate it into their workflows. Instead, experienced managers are better than juniors at developing prompting patterns that increase productivity. This suggests a surprisingly

---

[175] Anders, H., & Emilie, V. (2024). The Adoption of ChatGPT. *Working Paper*. https://static1.squarespace.com/static/5d35e72fcff15f0001b48fc2/t/668d08 608a0d4574b039bdea/1720518756159/chatgpt-full.pdf

[176] Naumovska, I., Gaba, V., & Greve, H. (2021). The Diffusion of Differences: A Review and Reorientation of 20 Years of Diffusion Research. *Academy of Management Annals, 15*, 377-405.

[177] Kellogg, K. C., Lifshitz, H., Randazzo, S., Mollick, E., Dell'Acqua, F., III, E. M., Candelon, F., & Lakhani, K. (2024). Don't Expect Juniors to Teach Senior Professionals to Use Generative AI: Emerging Technology Risks and Novice AI Risk Mitigation Tactics. *Working Paper*. https://papers.ssrn.com/sol3/papers.cfm?abstract_id=4857373

valuable role for some mid- and senior- managers as superusers who can support broader organization-wide adoption.

Superusers play a crucial role in knowledge dissemination, conducting demos, and educational efforts, discussing tool access, usage, best practices for personal productivity, integration with company tools, and compliance with corporate policies. Such practices have helped tech companies like Stripe, Google, Intercom, Zapier, and Adobe accelerate generative AI adoption.[178] The essence of broader adoption lies in knowledge sharing that fosters learning-by-doing, with superusers leading workshops, hackathons, or other practical events, sometimes supplemented by external experts or consultants. Forming tight-knit teams to undertake small projects benefits sub-departments by deepening employees' understanding of the technologies—this is the essence of learning-by-doing.

Superusers seem to become early adopters of new features and functions of generative AI as they attempt to rampantly apply technologies, often riding the wave of use and disuse that occurs. It is not necessarily that superusers are co-specialized with LLM functionalities.[179] They sometimes duplicate tasks that generative AI could perform as well—but a key is that they appear to be better able to select the best outputs, whether human or AI-generated, and learn from them.

---

[178] Tossell, B. (2024b). *Insights from Stripe, Google, Intercom, Zapier, Adobe and others and came up with 9 ways they help encourage employee adoption.* Februrary 1, 2024
https://x.com/bentossell/status/1753040606526394572
[179] Choudhary, V., Marchetti, A., Shrestha, Y. R., & Puranam, P. (2023). Human-AI ensembles. When can they work? *Journal of Management.*
https://journals.sagepub.com/doi/10.1177/01492063231194968

A prime example of generative AI adoption is Clearbit, a company recently acquired by Hubspot for $150M.[180] It prioritized use case exploration led by an AI champion, involving small, cross-functional teams and consultations with external experts, focusing on using AI agents to automate aspects of personal and work life. ne way to encourage widespread adoption is ensuring employees can see personal benefits beyond the value captured by the company, with superusers serving as perfect exemplars of this principle.

Experimentation: From Internal Tools to External Products. Once a significant portion of the workforce has adopted generative AI, the focus can shift to creating business value with it. The next phase entails experimentation with generative AI to create initiatives and programs. The fundamental challenge is the unpredictability of market outcomes. The goal should be to balance learning with short-term performance gains through learning-by-doing, ensuring that trials not only increase company profitability and growth but also develop routines, procedures, and structures that prepare the firm for future generative AI advancements.

A few archetypal market trials have proven to be effective in the early stages of developing larger capabilities. These projects rely on a basic level of technology adoption within the firm, with a large proportion of employees actively using generative AI. This frequently entails the extensive usage of generative AI for content development by individual employees or essential functions such as human resources, marketing, and IT.

---

[180] Tossell, B. (2024a). *How Clearbit uses AI - They were acquired for $150M by Hubspot last year.* January 25, 2024
https://x.com/bentossell/status/1750492229288779838

Chatbots: From Internal to External. After extensive use of generative AI in generating content, organizations become poised to embark on their first significant product development experiments with another architectural paradigm: chatbots. Chatbots, with a long history in internet-enabled services, span customer-facing and employee-facing applications across many industry verticals. Enhanced by LLM technology, chatbots have become more expressive and useful than their predecessors, which primarily relied on indexing and lookup tables for canned responses.

Generative AI frequently stimulates positive conversations about the potential of chatbots. However, it quickly becomes clear that developing generally applicable chatbots poses significant hurdles, including the possibility of inefficient execution or stakeholder rejection. While chatbots have the potential to add tremendous value, I believe they should be introduced gradually to reduce risks and promote learning. An optimal technique is to begin with a modest LLM-based chatbot experiment that causes low customer risk while maximizing learning opportunities.

Typically, this means focusing on an internal chatbot project that integrates information from knowledge management systems, such as corporate intranet databases, into an employee Q&A system. This "first experimental project" is ideal because the risk of failure has no direct impact on customers, the pace of development can match internal capabilities without external competitive pressure, and most organizations already have the necessary knowledge management systems with properly categorized information. Initial applications may involve answering questions about HR perks, onboarding, or accessing corporate history and competitive intelligence.

Developing this experiment allows for the refining of various activities, including the right internal marketing approach and data governance policies, while learning how to swiftly build generative AI products. Different use cases, each requiring unique user interfaces, can be incrementally deployed. A major technical risk is creating a brittle system that cannot be generalized or is prone to hallucinations or data exfiltration.[181] These technical challenges can be mitigated with further training or fine-tuning on a system initially exposed only to a subset of employees. As many of these learnings are applicable to external, customer-facing chatbots, this investment in knowledge is likely to yield a significant return.

Numerous emerging LLM chatbot technologies indicate that initial expectations for companies to train their own foundational models from scratch have shifted due to prohibitive costs. Instead, hybrid methods, such as Retrieval-Augmented Generation (RAG), have emerged. This framework allows a general model like ChatGPT to be augmented with texts retrieved from internal knowledge management systems, enabling the creation of answers based on combined representations of language and specific internal information.[182]

Other technical alternatives include OpenAI's GPTbuilder service, which allows the creation of custom GPT systems, including chatbots, based on uploaded knowledge

[181] Zoë Schiffer, C. N. (2023). *Amazon's Q has 'severe hallucinations' and leaks confidential data in public preview, employees warn.* Platformer. December 1, 2023 https://www.platformer.news/amazons-q-has-severe-hallucinations/
[182] Wolfe, C. R. (2024). *RAG is one of the best (and easiest) ways to specialize an LLM over your own data.* February 6, 2024 https://x.com/cwolferesearch/status/1754558231802769857

source files.[183] After these source files are uploaded, GPTs are created and hosted on OpenAI's servers. Many versions of these custom GPTs are now available, demonstrating the flexibility and potential for specialized applications.

Extending an internal chatbot incorporate access to various online information sources can deliver outside information as part of its results as well. The integration of internal and external knowledge can be highly powerful. Yet, a broader lesson may be that gaining the value of generative AI leverages the experience of working with external vendors in the generative AI ecosystem, learning to connect data sources and collaborate with external partners.

The development of an internal chatbot lays the groundwork for creating an external, customer-facing chatbot. Many of the technical methods remain the same, although the data and customer service orientation may differ. Typically, customer chatbots focus on service, troubleshooting, or providing information about products and services. What often changes is the elements of the customer journey and targeting, requiring sophisticated marketing processes. Articulating and communicating the value proposition for customers may require significantly more effort as customers have more heterogeneous and dynamic preferences.

External chatbots may find a natural place in call centers or with business process outsourcing (BPO) firms that manage customer relationships. Despite the ongoing issues of hallucinations and data exfiltration[184], with quality control, LLM

---

[183] Shipper, D. (2023). *How to build a chatbot in GPT-4.* April 18, 2023 https://x.com/danshipper/status/1648313206429949953

[184] Zoë Schiffer, C. N. (2023). *Amazon's Q has 'severe hallucinations' and leaks confidential data in public preview, employees warn.* Platformer.

chatbots trained on organization-specific data could significantly enhance productivity while reducing necessary headcount. The organizational structure of these functions could facilitate adoption, allowing chatbots to handle routine tasks and escalate more complex issues to human professionals, creating a synergistic blend of human and AI capabilities that can evolve as customer demands and technology capabilities change.

Organizations may conduct additional internal and external generative AI experiments to expand their capabilities, eventually moving on to product architectures such as co-pilot assistants or autonomous agents. The overarching goal is to build a strong capacity that can convert generative AI technology into commercial products and services while exploiting proprietary data and effectively addressing client needs. While some firms may choose to construct their own models, the majority will likely employ external suppliers for fundamental models, plugins, technology integration, or specialist AI data management.

Scaling Up: Redesigning Workflows and Business Model Innovation. Once organizations have product development capabilities that enable product experiments, they play out the results of these experiments until they find a success that can result in a substantial product improvement or new product line. Let us assume this product is gaining market traction and has found product market fit. Now, organizations face the critical challenge of scaling up operations and innovating their business model to effectively commercialize and capture value. This requires expanding

---

December 1, 2023 https://www.platformer.news/amazons-q-has-severe-hallucinations/

distribution channels, optimizing production capabilities, and ensuring that the infrastructure can handle increased demand.

Perhaps the most difficult challenge is analyzing and redesigning workflows in organizations to incorporate generative AI. Developing a one-off win with AI is easier than crafting a continuous stream of productivity improvements—this involves the redesign of how people work together to get the job done. One of the hardest elements is delegation of tasks to AI systems, as it can be unclear which tasks are most ready for replacement by AI.[185]

Then new tasks such as data sourcing and evaluation of outputs must be undertaken by humans. A human checker should be in the loop as the AI system may produce errors or inappropriate responses. This is particularly true with autonomous AI agents, which can generate catastrophic outcomes if errors go unchecked and propagate through the multi-agent systems.[186]

An important new task is managing the data that will be used to train generative AI. This involves establishing robust data warehouses and data lakes for efficient storage and retrieval. Data must be meticulously labeled and classified to ensure that AI models can learn from it accurately. Additionally, strong governance and security measures are required to protect sensitive information and comply with regulatory

---

[185] Rachel, W. (2024). *My recommendation to "learn AI" has become much more nuanced after teaching hundreds of people over the past 18 months.* 2024-06-26
https://twitter.com/rachel_l_woods/status/1805811212493496804
[186] Berrios, M. R. (2024). *Lessons from Parcha's Journey automating compliance workflows using AI and why autonomous agents aren't always the best solution.* Parcha's Resources. 2024-06-06
https://guidetoai.parcha.com/agents-arent-all-you-need/

requirements.[187] This includes ensuring data integrity and controlling access to prevent unauthorized usage. Additionally, the process involves storing data in LLMs and maintaining an updating function to keep the models current with the latest data inputs. This comprehensive approach ensures that generative AI systems are trained on high-quality, secure, and up-to-date data, enhancing their performance and reliability.

To grow their customer base, organizations must invest in marketing tactics that emphasize their generative AI product's unique value proposition, addressing both existing and unexplored markets. Furthermore, business model innovation is becoming increasingly vital as firms strive to effectively monetize their generative AI product. This entails rethinking pricing strategies, including subscription models, freemium services, and pay-per-use systems that are aligned with customer value perceptions and competitive environments. Companies must also evaluate and potentially pivot their value delivery mechanisms, employing data analytics to understand customer usage trends, feedback, and preferences in order to continuously improve their product and service offerings. Emphasizing customer success and support guarantees that as the product grows, user satisfaction remains high, building loyalty and boosting word-of-mouth marketing. Organizations can optimize the commercial value of their generative AI solutions by combining strategic scaling, process reform, business model innovation, and customer-centric methods.

The process emphasizes learning-by-doing, building generative AI organizational capabilities in stages, from

---

[187] Harney, M. (2024). *Good slide from Morgan Stanley on GenAI & data.* March 27, 2024 https://x.com/SaaSletter/status/1773000018024095997

identifying superusers and encouraging adoption to experimenting with products and scaling business operations.

Strategic Decision-making with Generative AI

The role of generative AI in aiding strategic decision-making at the highest levels of an organization has become a significant area of inquiry within strategic management and organizational science. Executives are interested in whether AI can contribute to the quality of strategic decisions, which are pivotal in enhancing organizational performance. One notable study has demonstrated that generative AI is particularly effective in evaluating alternative business models.[188] Interestingly, when the decisions of multiple LLMs are aggregated, their collective output can match or even surpass the expertise of human professionals. Generative AI has also shown promise in generating options within the well-known Blue Ocean Strategy framework, which emphasizes the creation of uncontested market spaces.[189] Given that this framework is highly conceptual, the ability of AI to generate meaningful and innovative options is a critical test of its language generation capabilities. The success of AI in this context highlights its potential to contribute to the development of groundbreaking strategies that can redefine market boundaries.

---

[188] Doshi, A. R. a. B., J. Jason and Mirzayev, Emil and Vanneste, Bart. (2024). Generative Artificial Intelligence and Evaluating Strategic Decisions. *Working Paper.* https://ssrn.com/abstract=4714776

[189] Olenick, M., & Zemsky, P. (2023). Can GenAI do strategy? *Harvard Business Review.* https://hbr.org/2023/11/can-genai-do-strategy

Another study focusing on entrepreneurs and investors found that generative AI can enhance cognitive processes underlying strategic decision-making, such as search, representation, and aggregation.[190] The study also noted that the impact of AI on performance is contingent on the speed at which AI develops and enables competitors to adopt similar technologies. The rapid pace of imitation and spillovers is a key determinant of whether competitive advantages derived from AI are sustainable.[191]

Some researchers have argued that the effectiveness of AI in strategic decision-making is heavily influenced by organizational design considerations.[192] Specifically, the type of design—whether AI and humans should specialize in different types of decisions or collaborate on the same decisions—can significantly affect outcomes. Moreover, the domain experience of humans using AI can play a dual role: while it can enhance performance by enabling users to better complement algorithmic advice, it can also hinder performance if experienced users are overly averse to accurate algorithmic recommendations.[193]

Given organizations might integrate AI differently, certain organizational processes may be more conducive to

---

[190] Csaszar, F. A., Ketkar, H., & Kim, H. (2024). Artificial Intelligence and Strategic Decision-Making: Evidence from Entrepreneurs and Investors. *Working Paper*.
https://papers.ssrn.com/sol3/papers.cfm?abstract_id=4913363
[191] Davis, J. P., & Aggarwal, V. A. (2020). Knowledge mobilization in the face of imitation: Microfoundations of knowledge aggregation and firm-level innovation. *Strategic Management Journal, 41*(11), 1983-2014.
[192] Puranam, P. (2021). Human–AI collaborative decision-making as an organization design problem. *Journal of Organization Design, 10*(2), 75-80.
[193] Allen, R., & Choudhury, P. R. (2022). Algorithm-augmented work and domain experience: The countervailing forces of ability and aversion. *Organization Science, 33*(1), 149-169.

generating different types of content and decisions: iterative approaches, where outputs are refined through a back-and-forth exchange between AI and human inputs, may lead to more innovative and effective strategic decisions.[194] Many organizational processes mirror computational arrangements found in computer science, leading some to conceptualize organizations themselves as artificial intelligences.[195]

However, a counterpoint in this debate is the argument that AI cannot fully replace humans in strategic decision-making because such decisions often involve causal logic and reasoning.[196] While machine learning and AI can be employed to build theories through an inductive process from data, strategic decision-making also requires deductive reasoning, which is integral to scientific rigor. Additionally, LLMs are often biased by their limited training data,[197] and may therefore find it difficult to explore logical chains of reasoning if they are not represented in prior history.

If AI is incapable of engaging in deductive logic, it may fall short of the requirements for a more scientific approach to strategic decision-making.[198] Research has shown that LLMs do not inherently reason; instead, efforts must be made to structure

---

[194] Raisch, S., & Fomina, K. (2024). Combining human and artificial intelligence: Hybrid problem-solving in organizations. *Academy of Management Review.* https://doi.org/10.5465/amr.2021.0421

[195] Csaszar, F. A., & Steinberger, T. (2022). Organizations as artificial intelligences: The use of artificial intelligence analogies in organization theory. *Academy of Management Annals, 16*(1), 1-37.

[196] Felin, T., & Holweg, M. (2024). Theory is all you need: AI, human cognition, and decision making. https://ssrn.com/abstract=4737265

[197] Cowgill, B. (2019). Bias and productivity in humans and machines. https://ssrn.com/abstract=3433737

[198] Camuffo, A., Cordova, A., Gambardella, A., & Spina, C. (2020). A scientific approach to entrepreneurial decision making: Evidence from a randomized control trial. *Management Science, 66*(2), 564-586.

processes and extract reasoning to enable them to engage in causal logic.[199] This remains an active field of computer science research, with anticipated advancements that could significantly enhance the role of AI in strategic decision-making.

One area of future application may be in the use of generative AI to simulate conditions that are too complex to think through. Simulations have been a good tool for strategic decision-making, particularly in war games. A recent system uses LLMs to model China's approaches to war and other international conflicts under diverse settings.[200] The model enables decision-makers to consider counterfactuals for which there is very little historical data. LLM-based simulations may be particularly useful in modeling customer behavior under different market, labor, pricing, and other conditions that inform economic outcomes.[201] Such models have become increasingly advanced world-models—for instance, video image models can perform effectively as world simulators.[202] That is, these models have the capability of generating entire online worlds. These simulations can become more effective tools in decision-making with innovations in LLMs.

---

[199] Huang, J., & Chang, K. C. C. (2023). Towards reasoning in large language models: A survey. https://arxiv.org/abs/2212.10403

[200] Hua, W., Fan, L., Li, L., Mei, K., Ji, J., Ge, Y., Hemphill, L., & Zhang, Y. (2024). War and peace (WarAgent): Large language model-based multi-agent simulation of world wars. https://arxiv.org/abs/2311.17227

[201] Horton, J. J. (2023). Large language models as simulated economic agents: What can we learn from Homo Silicus? https://arxiv.org/abs/2301.07543

[202] Brooks, T., Peebles, B., Homes, C., DePue, W., Guo, Y., Jing, L., Schnurr, D., Taylor, J., Luhman, T., Luhman, E., Ng, C., Wang, R., & Ramesh, A. (2024). Video generation models as world simulators. *Working Paper.* https://openai.com/research/video-generation-models-as-world-simulators

Chapter 7

—

Industry Lens: Business Opportunities and Disruption Avoidance

"AI will be the most transformative technology of the 21st century. It will affect every industry and aspect of our lives." Jensen Huang, CEO at NVIDIA[203]

Understanding the applications of generative AI at the industry level provides is a helpful perspective. Business practices within industries tend to be similar, which means that identifying best practices can have wide-reaching implications across them. As outlined earlier, generative AI is poised to continuously evolve, resulting in substantial shifts and changes in how business is done. Scholars often regard generative AI as a "General Purpose Technology" that impacts numerous, if not all, industries in various ways,[204] like how the internet and mobile applications have in recent times.

---

[203] Martin, D. (2024). 6 Bold Statements By Nvidia CEO Jensen Huang On AI's Future. *CRN Channel Co.* August 5, 2024

[204] Bresnahan, T. F., & Tratjenberg, M. (1995). General Purpose Technologies: 'Engines of Growth'? *Journal of Econometrics, 65*(1), 83-108.

Considering the historical example of electricity helps illustrate the concept of General Purpose Technologies more clearly. [205] The electrification of the global economy around the turn of the 19th century revolutionized industries that had previously relied on steam engines for power. The initial adoption of electricity was driven by straightforward benefits, such as energy cost savings. However, it was later discovered that reengineering entire factories could significantly amplify efficiency gains. This often involved a complete redesign of the factory floor to leverage electricity's advantages. By the 1930s, innovative companies of that era capitalized on this technological shift. [206]

Many economic historians believe that these profound disruptions resulted from how organizations adapted—or failed to adapt—to electricity. As a General Purpose Technology, electricity impacted almost every business across various sectors, although the speed and magnitude of its influence varied by industry. Developing a competitive edge with this new technology depended on identifying the best practice use cases of electricity tailored to particular businesses.

Similarly, developing a competitive advantage with generative AI may hinge on identifying industry-specific best practices. Industries such as consulting, software, media, and entertainment—predominantly knowledge-based—might be most affected by generative AI. However, since most businesses incorporate some form of knowledge work, no industry will remain unaffected. The key lies in understanding and utilizing

---

[205] Brynjolfsson, E., & Mcafee, A. (2017). *Machine, Platform, Crowd: Harnessing Our Digital Future.*

[206] Goldfarb, B. (2005). Diffusion of general-purpose technologies: understanding patterns in the electrification of US Manufacturing 1880-1930. *Industrial and Corporate Change, 14*(5), 745-773.

the best practice use cases for each industry. Below, I describe some emerging best practices from several prominent industries.

Financial Services

Banking (Commercial and Retail)

Customer Onboarding. Artificial intelligence improves productivity and streamlines onboarding by automating form filling, replying to concerns regarding banking protocols, and managing the complexities of opening new accounts. It also ensures the accuracy of customer data and identity verification, while individualized onboarding experiences address specific consumer preferences and concerns. Generative AI's role includes detecting and resolving any differences early in the process, making transfers for new clients easier and more welcome.

Customer Support. Customer support has become more responsive, with AI systems handling inquiries and resolving common issues swiftly. AI enables banks to provide quick responses while maintaining high customer satisfaction levels. AI-powered chatbots and virtual assistants can interpret and respond to client requests in real time, either providing accurate and relevant information or seamlessly escalating complex situations to human agents.

Loan Origination. Generative AI makes it easier to acquire and verify applicant information, assess eligibility requirements, and automate decision-making procedures. This accelerates loan approvals and improves the accuracy of risk evaluations, resulting in better informed lending decisions.

Fraud Detection. Generative AI excels in detecting and mitigating fraudulent activity by analyzing transaction patterns. Its ability to learn and adapt to new fraudulent methods guarantees that banks effectively secure their customers' assets. AI systems analyze transactions in real time, identifying irregularities that may suggest fraud and automatically taking preventative steps.

Credit Scoring. AI is transforming credit scoring by utilizing sophisticated algorithms to assess the creditworthiness of individuals or corporations. By analyzing large datasets, AI can find nuanced patterns and factors that influence credit risk, providing a more thorough evaluation than traditional techniques. This results in fairer and more accurate credit determinations. According to a recent McKinsey and Company survey, generative AI is expected to have a significant impact on the financial industry, with 80% intending to adopt gen AI technology for this purpose by 2024.[207]

Customized Financial Advice. Providing tailored financial guidance is now more achievable with generative AI. By analyzing customers' financial histories and objectives, AI can deliver personalized advice, helping customers make informed decisions. It can suggest strategies for savings, investments, and managing debt, aligning recommendations with each customer's unique financial situation.

Compliance. AI aids in ensuring regulatory compliance by automating the monitoring of transactions and activities. Generative AI can detect deviations from regulatory standards, reducing the risk of non-compliance. It also

---

[207] Kremer, A., Govindarajan, A., Singh, H., Kristensen, I., & Li, E. (2024). Embracing generative AI in credit risk. *Mckinsey and Company.* https://www.mckinsey.com/capabilities/risk-and-resilience/our-insights/embracing-generative-ai-in-credit-risk

streamlines reporting processes, making it easier for banks to adhere to laws and regulations.

Financial Planning. AI assists individuals and businesses in planning their financial futures by providing tools for budgeting, forecasting, and scenario analysis. Generative AI takes into account various financial goals and market conditions, offering strategies that are both realistic and optimized for future growth.

KYC and AML. AI also enhances know your customer (KYC) and anti-money laundering (AML) efforts. Automated checks and verifications speed up onboarding, while maintaining stringent security measures. AI-driven systems can sift through large volumes of data to identify potential risks efficiently.

Risk Management. Generative AI can help analyze different risks, such as credit, market, and operational, and suggest strategies to mitigate them. AI can forecast potential vulnerabilities and recommend preventative actions by processing complex datasets, aiding banks in safeguarding against unforeseen events. Financial institutions expect to implement generative AI across the risk management lifecycle, from credit applications to portfolio monitoring to collections.[208]

Process Automation. From creating notes on employee retention credits to incorporating best practices around gamification of deposits, AI is making banking operations more inclusive and accessible. It is also playing a role in helping redesign webpages for accessibility, ensuring that all

---

[208] Kremer, A., Govindarajan, A., Singh, H., Kristensen, I., & Li, E. (2024). Embracing generative AI in credit risk. *Mckinsey and Company.* https://www.mckinsey.com/capabilities/risk-and-resilience/our-insights/embracing-generative-ai-in-credit-risk

customers, regardless of ability, have equal access to banking services.

Wealth Management

Personalized Investment Strategies and Robo-Advisors. Automated financial planning services provided by robo-advisors are becoming increasingly sophisticated, thanks to generative AI. These advisors offer personalized investment advice with minimal human supervision, adjusting strategies based on real-time market data and individual financial situations. AI customizes investing strategies to meet individual risk tolerances, financial objectives, and time horizons. AI can create strategies that maximise returns while reducing risk by analysing historical data and current market trends.

Market Analysis and Trend Prediction. Using AI for market analysis and trend prediction enables investors to make better judgments. AI systems sift through massive volumes of market data to uncover possible investment opportunities and predict future trends.

Portfolio Management and Optimization. Continual monitoring and optimization of investment portfolios is facilitated by AI. It rebalances portfolios to align with clients' objectives, ensuring that investments are strategically positioned for growth and risk management.

Investment Banking

JP Morgan's Internal ChatGPT. An AI investment advisor, like JP Morgan's Internal ChatGPT, enhances decision-making and advisory services in investment banking. By providing deep insights into market trends and investment opportunities, AI supports bankers in making strategic decisions.

Deal Origination and Structuring. AI identifies potential deal opportunities and optimizes deal structures based on market data and trends. This only accelerates the deal origination process and ensures structures are optimized for success.

Financial Modeling and Valuation. Generative AI can help create more accurate and complicated financial models for valuation, deal analysis, and scenario planning. These models enable more exact evaluations and strategic planning. For example, in mergers and acquisitions, AI can examine and predict potential synergies, assisting with strategic decision-making. This skill ensures that transactions are more likely to succeed and add value.

Client Relationship Management. AI analyzes client data to provide insights into client needs and opportunities. This improves client relationships and also enables the delivery of more targeted, customized, and effective financial advice.

Tax Management. AI can also aid in the complex and consequential area of tax management. One notable example is TaxGPT, a specialist product. TaxGPT streamlines tax filing and provides individuals and businesses with a stress-free alternative for tax preparation and consultation. This AI-powered application guides users through the complexity of tax regulations and identifies possibilities to lower tax liabilities.

Generative AI for Enhanced UX/UI. The revamping of webpages for greater accessibility and the generation of inclusive content are two examples of how generative AI is improving the user experience in financial services. AI improves digital platforms by evaluating feedback and statistics to better fulfill the demands of users.

Content Creation and Management. AI has made writing and optimizing text for bank websites, job postings, and promotional materials more efficient. By properly targeting certain audiences, banks may ensure that their messages reach their targeted audience and increase interaction.

E-commerce

A large chunk of global commerce has shifted online. As of 2024, nearly 20% of the world's commerce takes place on the internet, representing over $6.3 trillion in value and expanding at an annual rate of about 20%.[209] In this rapidly growing digital marketplace, generative AI has begun playing a crucial role.

Advertising. AI plays an important part in developing more effective e-commerce advertising tactics. By evaluating client data, AI enables the creation of customized ad campaigns that appeal to certain groups, making marketing efforts more precise and cost-effective. AI also excels at dynamic content generation, automatically developing and testing various advertising content to see which performs better across different audience segments. Furthermore, real-time bidding optimization using AI algorithms optimizes ROI on ad spend by making more informed bidding decisions in ad placements.

Information Organization and Generation. E-commerce sites benefit greatly from AI's capacity to generate compelling and SEO-friendly product descriptions and captions, which make things more appealing and simpler to find online. AI also simplifies the building of e-commerce websites by providing personalized, optimized layouts and product

---

[209] Chevalier, S. (2024). *Retail e-commerce sales worldwide from 2014 to 2027*. Statista. https://www.statista.com/statistics/379046/worldwide-retail-e-commerce-sales/

placements that improve user experience and revenue potential. Furthermore, AI monitors consumer behavior and site interactions to enhance product placement and suggestions, guaranteeing that customers discover things they are more likely to buy.

Better Relationships with Suppliers. AI technology is becoming an important tool for improving supplier relationships. AI-powered negotiating tools utilize deep learning to monitor market dynamics and help businesses negotiate better contracts. Walmart's use of AI to boost its bargaining skills is a case in point. Some reports imply that agreements are being completed faster, allowing the corporation to focus on larger contracts.[210]

Customer Relationships. AI significantly improves customer service in e-commerce settings. AI-powered chatbots handle inquiries effectively, provide support, and engage customers, improving customer experience.[211] AI applications also boost engagement through personalized interactions and proactive customer services, making the shopping experience more personalized and responsive.[212] In addition, AI is used to efficiently manage and resolve customer disputes, minimizing the need for extensive human intervention.

Additional Potential Applications. AI's capabilities extend to managing inventory by predicting needs

---

[210] Kourosh, S. (2023b). *"Recently, it has been reported that AI bots are negotiating terms and closing deals with vendors for WMT. .* 2023-06-08 https://twitter.com/kouroshshafi/status/1666589434123534339

[211] Tyler Hutcherson, H. C. (2023). Build an E-commerce Chatbot With Redis, LangChain, and OpenAI. https://redis.com/blog/build-ecommerce-chatbot-with-redis/

[212] Esther, S. (2023). *Tutorial: How to build an e-commerce chatbot using #OpenAI, @Redisinc, and @LangChainAI* 2023-04-13 https://twitter.com/estherschindler/status/1646191037717839873

and optimizing stock levels, hence reducing both overstock and understock issues. Price optimization algorithms alter prices dynamically based on market demand, competition, and inventory levels, making pricing strategies as effective as feasible. AI-powered consumer sentiment analysis enables organizations to evaluate social media reviews and feedback to gauge customer sentiment, allowing for proactive brand reputation management. Furthermore, AI systems are effective in detecting and preventing fraudulent transactions by analyzing transaction patterns and identifying anomalies. Finally, visual search and recognition technologies allow users to search for products using photographs, which improves the shopping experience by suggesting similar things.

Consulting

The adoption of generative AI in consulting is driven mostly by its capacity to cut costs and improve the application of expertise to client's problems. Generative AI reduces labor costs by automating jobs previously handled by highly compensated professionals. It enables the efficient application of deep information via intuitive, language-based interfaces, improving recommendations, interactions, and presentations while expediting various consulting processes.

Generative AI enhances the clarity and impact of written content in reports, presentations, and client engagements. It automates the development of PowerPoint slides by translating summarized material and data insights into visually appealing presentations. It can also produce bespoke interview questions based on unique consulting scenarios, such as client or project demands. Furthermore, AI recognizes and recommends tools and technology that help improve consulting methods and client outcomes.

McKinsey assisted ING in deploying a bespoke AI chatbot, which is a prime example of generative AI in a consulting project. This effort aims to improve customer service by responding to inquiries quickly and accurately, showcasing the practical application of artificial intelligence in enhancing client interactions. Developed in conjunction with QuantumBlack, the AI chatbot provided specialized support far faster than past solutions, increasing customer satisfaction and operational efficiency. This initiative established ING as a pioneer in applying generative AI in the banking industry.[213]

Use Cases from Consulting Practice

- ***Strategy Formulation and Execution:*** Generative AI aids consultants by crafting strategies and operational tactics tailored to specific market conditions and client problems. It analyzes vast amounts of data to provide strategic insights that are deeply aligned with client goals.
- ***Due Diligence and Risk Analysis:*** AI automates the labor-intensive processes of data collection and analysis during due diligence, significantly reducing the time required and increasing the accuracy of risk assessments.

---

[213] Miglio, A. D., Giovine, C., Hauser, S., Ouass, M., & Wildt, N. V. d. (2024). *Banking on innovation: How ING uses generative AI to put people first.* Mckinsey and Company. https://www.mckinsey.com/industries/financial-services/how-we-help-clients/banking-on-innovation-how-ing-uses-generative-ai-to-put-people-first

- ***Implementation Support:*** Generative AI provides automated guidance for implementing complex systems and processes. By simulating different implementation scenarios, AI helps in predicting outcomes and reducing errors, thereby speeding up the implementation phase.
- ***Training and Change Management:*** AI-driven training modules and simulations prepare clients and their teams for transitions, especially in adopting new technologies. These tools are designed to enhance learning efficiency and ensure the smooth adoption of new systems.

Generative AI also plays a crucial role in optimizing business processes, developing personalized client solutions, and advising on digital transformation strategies. Taken together, its evolving role in consulting promises to reshape the industry, driving innovation and enhancing the value delivered to clients. The technology will assume greater importance in the industry as it continues to mature, which means the consulting industry will need to continually adapt.

Legal Profession

Generative AI is poised to deeply impact knowledge-work-based industries, such as law. The legal profession is a prominent example of a knowledge-based sector, with most of its work involving content generation. In fact, the promise of AI has been well known in the legal industry for a long time.[214]

---

[214] Marwaha, A. (2017). *7 Ways artificial intelligence can benefit your law firm.* American Bar Association.
https://www.americanbar.org/news/abanews/publications/youraba/2017/september-2017/7-ways-artificial-intelligence-can-benefit-your-law-firm/

Recently, however, the technology's integration into legal services promises enhancements in efficiency, accuracy, and client relations.

For Clients: Engaging with a Lawyer

When generative AI first emerged, it became clear that LLMs could master various exams necessary to become an attorney[215]— but could generative AI help with the process of engaging legal services for actual clients? Indeed, generative AI has been shown to improve several stages of the legal process. For example, providing information about qualified lawyers' credentials and service quality improves the search process. Clients can use AI algorithms to discover trustworthy lawyers and avoid frauds by evaluating lawyer reviews and ratings. Furthermore, generative AI can help clients navigate the mechanics of meeting planning, such as directions and appointment scheduling, making the process more user-friendly and accessible.

For Law Firms

Generative AI reshapes the workflow within law firms by automating many tasks traditionally handled by associates. This includes:

- **Research and Document Creation.** AI enhances legal research and document drafting efficiency. It can autonomously generate legal documents such as contracts and briefs, review documents for accuracy, and answer common legal questions. This automation extends to tasks like contract drafting and

---

[215] Katz, D. M., Bommarito, M. J., Gao, S., & Arredondo, P. (2023). GPT-4 passes the Bar Exam. https://ssrn.com/abstract=4389233

comprehensive document reviews, streamlining processes that were once time-consuming.

- **_Preparation for Cases and Trials._** AI assists in case preparation by uncovering relevant precedents, verifying citation formats, defining applicable statutes and regulations, and crafting discovery questions. This support helps legal teams prepare more thoroughly and efficiently for court proceedings.

- **_Advanced Prompting Techniques._** Generative AI performs well when guided by advanced prompting mechanisms. For example, telling the AI to "Act as my opposing counsel" can give lawyers with robust simulations of probable trial arguments, so improving their readiness. Providing clear settings and desired objectives can help the AI produce more relevant and realistic results for certain circumstances.

- **_Client Management:_** In client interactions, generative AI proves valuable by summarizing complex legal cases and concepts in plain language, enhancing client understanding and engagement. It also aids in creating legal marketing content that communicates a firm's expertise effectively to potential clients.

Risks Associated with Generative AI in Legal Applications

While generative AI has various benefits in legal practice, it also carries substantial hazards, notably in terms of the veracity of created information. The possibility of hallucination—AI producing inaccurate or misleading information—is especially troubling in the legal area, where the stakes are enormous. Errors or inaccuracies in legal papers or advice can have major effects in court rulings and may result in liability for the law practice.

To reduce these hazards, comprehensive examinations and rechecks by qualified personnel are required. The usage of specific generative AI tools, such as Harvey, which incorporates a large collection of legal papers and case law, can help increase the accuracy of AI-generated legal content.

Overall Assessment: Impact of Generative AI on Legal Efficiency and Quality

Recent research and implementations have revealed that generative AI has a significant impact on the efficiency and quality of legal work. A significant example from New Zealand contrasted the performance of law experts to advanced language models such as GPT-4 and others in legal document evaluation tasks. The findings revealed that language models could complete reviews in seconds, saving 99.97% of the time required by traditional approaches. Furthermore, these models run at a fraction of the expense of human legal practitioners. Senior lawyers who examined the work produced by these models discovered that the quality of legal outcomes was frequently higher when AI was engaged. This demonstrates the potential for generative AI to not only speed legal processes but also improve the quality of legal outputs.[216]

However, in other studies the impact of generative AI on the legal profession has been mixed or even negative. A prominent study from Stanford university found that generative AI gave correct answers only 20-65% of the time, with

---

[216] Lauren Martin, N. W., Stephanie Yiu, Lizzie Catterson, Rivindu Perera. (2023). Better Call GPT, Comparing Large Language Models Against Lawyers. *Working Paper.* https://arxiv.org/html/2401.16212v1

hallucinations rates of 17-33%.[217] There is some indication that new legal generative AI solutions may already beat these more pessimistic benchmarks—whatever happens, it seems clear that these solutions will only improve with better training data, algorithms, and fine-tuning.

Taken together, these findings highlight generative AI's transformational potential for altering legal practices, making them more efficient and potentially correct. However, the integration of these technologies must be approached with prudence, with experienced legal experts providing rigorous monitoring and verification to protect against the inherent hazards of AI-generated errors.

Academia and Education

Generative AI is set to profoundly influence academia and education, sectors fundamentally involved with the retention, generation, and dissemination of knowledge.

For Educators

Writing. Generative AI has changed interactions with academic literature. Researchers can now easily locate texts, summarize them, and engage in "conversations" with a corpus of documents, thus enhancing accessibility and interaction with academic content. This capability is especially beneficial for exploring new or obscure areas, providing researchers with a quick means to familiarize themselves with new domains. Apps like Perplexity, valued for their ability to preserve citations, are

---

[217] Ziniti, C. (2024). *Stanford's working on a study showing most AIs fail at legal work.* . June 6, 2024

integral to this new approach.[218] [219] AI also assists in managing and drafting manuscripts by enhancing citation management and converting references to different formats. Additionally, specialized tools like Paperpal are emerging to specifically cater to academic writing.

Scientific Research. AI significantly aids in managing surveys, data analysis, and especially data collection from unstructured or complex sources. Simon Willison's Datasette Extract, for instance, uses GPT-4 to transform unstructured data into structured database entries, highlighting the practical utility of AI in data management.[220] Moreover, as mathematician Terrence Tao suggests, generative AI is becoming a reliable collaborator in research, capable of performing logical deductions and assisting in mathematical proofs.[221]

Teaching. Generative AI significantly aids knowledge dissemination. For example, Ethan and Lilach Mollick suggest that AI can fulfill various roles such as AI-tutor, AI-coach, and AI-mentor, each designed to enhance different aspects of learning:[222] As they describe, these can be summarized as, "AI-

---

[218] Bilal, M. (2023). *Thanks to ChatGPT, hundreds of AI apps are being released every week now.* April 15, 2023

[219] Cowen, T., & Shipper, D. (2024). *Economist Tyler Cowen on How ChatGPT Is Changing Your Job - Ep. 7 with Tyler Cowen.* January 24, 2024 https://youtu.be/5JZtPE8LU-4?si=bzayvXEBwPmI3MlA

[220] Willison, S. (2024). *Datasette Extract is a new Datasette plugin that uses GPT-4 (and the new GPT-4 Vision) to extract structured data from unstructured text and images and insert it into a SQLite database table.* April 10, 2024 https://x.com/simonw/status/1777820654487925245

[221] Tao, T. (2024). Embracing change and resetting expectations. *AI Anthology.* https://unlocked.microsoft.com/ai-anthology/terence-tao/

[222] Mollick, E., & Mollick, L. (2023). Assigning AI: Seven Approaches for Students, with Prompts [http://dx.doi.org/10.2139/ssrn.4475995]. *Working Paper.*

tutor, for increasing knowledge, AI-coach for increasing metacognition, AI-mentor to provide balanced, ongoing feedback, AI-teammate to increase collaborative intelligence, AI-tool for extending student performance, AI-simulator to help with practice, and AI-student to check for understanding."[223] The integration of AI in teaching highlights the importance of maintaining a "human in the loop" to ensure genuine learning and skill development.

For Students

Many students make use of generative AI for document creation, learning enhancement, and summarization. However, this raises debates about academic integrity and the nature of learning, questioning whether AI serves as an accelerant or a substitute for their studies. Tyler Cowan suggested that has a particularly large impact on what happens outside of the instructor's eye, suggesting that the "age of homework is over."[224] Ethan Mollick also discusses the significant impacts of AI outside direct supervision and the ensuing "homework apocalypse," suggesting a shift toward more monitored testing methods.[225] In response, some institutions like the University of Michigan are embracing AI, offering custom AI platforms to

---

[223] Mollick, E., & Mollick, L. (2023). Assigning AI: Seven Approaches for Students, with Prompts [http://dx.doi.org/10.2139/ssrn.4475995]. *Working Paper.*
[224] Cowen, T., & Shipper, D. (2024). *Economist Tyler Cowen on How ChatGPT Is Changing Your Job - Ep. 7 with Tyler Cowen.* January 24, 2024 https://youtu.be/5JZtPE8LU-4?si=bzayvXEBwPmI3MlA
[225] Mollick, E. (2023a). *The Homework Apocalypse.* July 1, 2023 https://www.oneusefulthing.org/p/the-homework-apocalypse

their entire community, signifying a trend in higher education to integrate these technologies.[226]

The impact of generative AI across academia and education is profound, presenting both opportunities and challenges that necessitate careful navigation to harness its full potential effectively.

Medicine

Generative AI has the potential to change the field of medicine, which relies largely on skill and knowledge. Its ability to integrate and reason through large amounts of medical data makes it a valuable asset in improving healthcare delivery and research.

Diagnosis and Prediction

Generative AI, such as LLMs, can store and process large amounts of information regarding symptoms and diseases. This feature enables them to provide stronger differential diagnoses by combining large volumes of medical data. Generative AI is very effective at diagnosing undiscovered diseases, delivering insights where human doctors may fail. A study found that LLMs properly identified 59% of difficult situations, compared to 33% for human doctors.[227] Although hallucinations may be a

---

[226] Burns, T. (2023). *ITS debuts custom artificial intelligence services across U-M*. August 21, 2023 https://record.umich.edu/articles/its-debuts-customized-ai-services-to-u-m-community/

[227] Daniel McDuff, M. S., Tao Tu, Anil Palepu, Amy Wang, Jake Garrison ,Karan Singhal, Yash Sharma, Shekoofeh Azizi, Kavita Kulkarni, Le Hou, Yong Cheng, Yun Liu, S Sara Mahdavi, Sushant Prakash, Anupam Pathak, Christopher Semturs, Shwetak Patel, Dale R Webster, Ewa Dominowska, Juraj Gottweis, Joelle Barral, Katherine Chou, Greg S Corrado, Yossi Matias, Jake Sunshine, Alan Karthikesalingam, Vivek Natarajan. (2023). Towards

particularly big problem in medicine, an additional AI instance can be deployed to identify and correct errors and hallucinations in AI outputs, enhancing the reliability of medical AI applications.

A more futuristic possibility is that generative AI may be able to not only diagnose current conditions but predict future ones. A new generative AI model called "Med-Gemini-Polygenic" appears to do exactly that.[228] Using genomic data, it computes polygenic risk scores for various conditions such as depression, stroke, coronary disease, and diabetes. It appears to beat the current best PRS models on diseases that are both in and out of the distribution.

Accurate Differential Diagnosis with Large Language Models. *Working Paper*. https://arxiv.org/pdf/2312.00164

[228] Azizi, S. (2024). *Med-Gemini-Polygenic is the first LMM to predict health outcomes from genomic data*. May 7, 2024 https://x.com/AziziShekoofeh/status/1787657316071780548

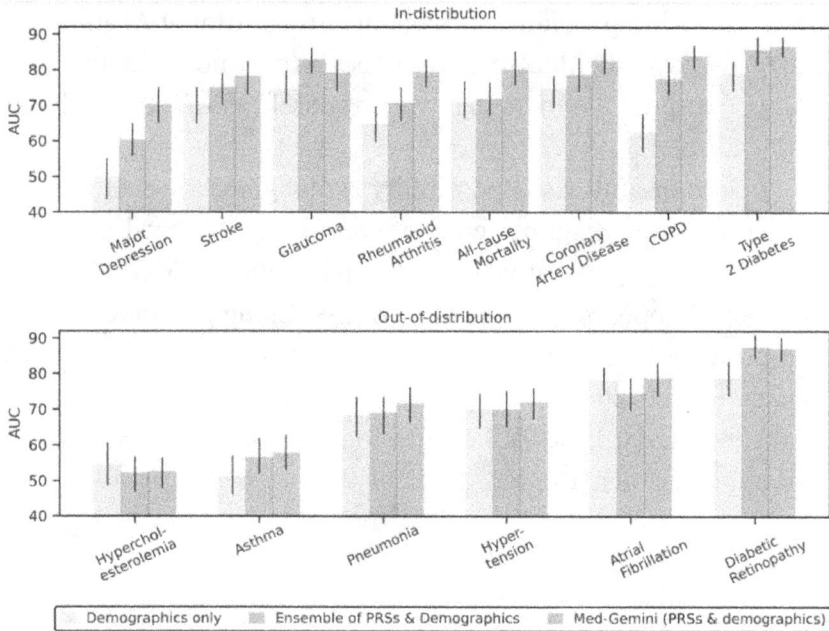

This model is part of a family of medical models—Med-Gemini—that leverage Gemini's basic language abilities, but are trained on the latest medical knowledge to perform many tasks in medicine.[229] An example includes analyzing chest X-rays for disease, including putting outputs in various modalities such as pathology, ophthalmology, radiology, and dermatology. Such advanced and customized medical models are likely to have a large impact on medicine.[230]

One challenge with AI-based diagnosis, though, is that it can sometimes be in conflict with how doctors usually wish to

---

[229] Banks, A. (2024). *Conversation with Med-Gemini for medical tasks.* May 1, 2024 https://x.com/thealexbanks/status/1785654353308581947
[230] Azizi, S. (2024). *Med-Gemini-Polygenic is the first LMM to predict health outcomes from genomic data.* May 7, 2024
https://x.com/AziziShekoofeh/status/1787657316071780548

understand the logic behind the diagnosis before acting on it.[231] AI results are often opaque so doctors must establish trust in these systems—or the systems must increase the explanatory power behind their predictions—before doctors are willing to use them. Yet the research on how to do so is not clear, with some research even indicating that increasing transparency into what the algorithms do and how to interpret them may lead to less effective use as it biases humans to be too overconfident in AI use.[232]

Enhancing Doctor-patient Encounters

AI can improve interactions between doctors and patients by offering more empathetic communication and being less susceptible to fatigue.[233] AI could potentially be more persuasive than human doctors in convincing patients about treatment plans. It can also simulate various conversational scenarios, such as a child talking to his or her mother, to assist doctors in understanding emotional and situational dynamics.

Operational Improvements

AI is useful for eliminating waste, detecting fraud, improving reimbursement processes, and minimizing total

---

[231] Lebovitz, S., Lifshitz-Assaf, H., & Levina, N. (2022). To engage or not to engage with AI for critical judgments: How professionals deal with opacity when using AI for medical diagnosis. *Organization Science, 33*(1), 126-148.
[232] DeStefano, T., Kellogg, K. C., Menietti, M., & Vendraminelli, L. (2022). Why providing humans with interpretable algorithms may, counterintuitively, lead to lower decision-making performance. https://ssrn.com/abstract=4246077
[233] John W. Ayers, A. P., Mark Dredze, Eric C. Leas, Zechariah Zhu, Jessica B. Kelley, Dennis J. Faix, Aaron M. Goodman, Christopher A. Longhurst, Michael Hogarth, Davey M. Smith. (2023). Comparing Physician and Artificial Intelligence Chatbot Responses to Patient Questions Posted to a Public Social Media Forum. *JAMA Internal Medicine, 183*(6), 5890596.

costs. It also automates and personalizes patient communication, resulting in constant follow-ups and information sharing. It also helps to verify the legality of medical actions and streamlines billing processes to improve compliance and efficiency. As it stands, medical professionals frequently struggle to incorporate generative AI outputs into their decision-making due to a lack of transparency about how the findings are generated and their appropriateness. Over time, we should hope for advances in how this knowledge is integrated into their practice and medical decision-making to benefit from the capabilities of generative AI.[234]

Clinical Trials and Research

AI optimizes the design and management of clinical trials, accelerating the development of new treatments and cures. It also tailors explanations of scientific findings to different audiences, enhancing the dissemination of complex medical knowledge.

Medical Education

AI's capabilities in analyzing medical text and offering diagnoses were demonstrated in a recent Stanford study published in JAMA, indicating a significant role for AI in medical education and training.[235] Known as "DoctorGPT," this specific LLM has shown the ability to pass the US Medical

---

[234] Lebovitz, S., Levina, N., & Lifshitz-Assaf, H. (2021). Is AI ground truth really true? The dangers of training and evaluating ai tools based on experts' know-what. *MIS Quarterly*, *45*(3), 1501-1525.

[235] Eric Strong, A. D., Yingjie Weng, Andre Kumar, MD, Poonam Hosamani, Jason Hom, Jonathan H. Chen. (2023). Chatbot vs Medical Student Performance on Free-Response Clinical Reasoning Examinations. *JAMA Internal Medicine*, *183*(9), 1028-1030.

Licensing Exam, functioning offline and across different platforms while keeping health data private.[236]

Empowering Patients

In modern times, patients often visit doctors after having done Google searches. But the information gained from an internet search is not appropriately contextualized, and can cause more harm than good. Generative AI provides balanced and contextual information, improving access to and equity of medical knowledge, filling gaps in care, and supporting behavioral changes.[237]

Staffing and Productivity Enhancements

AI helps reduce doctor shortages and exhaustion by taking over routine activities and administrative responsibilities, allowing doctors to focus more on patient care. AI solutions can give performance statistics, identifying areas for improvement and assuring high-quality medical care. Finally, AI can help with data administration in health care by creating patient records in standardized formats such as HL7 FHIR, ensuring interoperability and ease of data management.

The broader implications of generative AI for the medical industry include prospective changes in healthcare provider responsibilities as well as ethical problems. The growing trio of patient, AI, and doctor is becoming a model for future healthcare delivery, emphasizing the revolutionary

---

[236] Roemmele, B. (2023a). *DoctorGPT is an LLM that can pass the US Medical Licensing Exam.* August 14, 2023
https://x.com/BrianRoemmele/status/1691076100498276352
[237] Lee, P., Goldberg, C., & Kohane, I. (2023). *The AI Revolution in Medicine: GPT-4 and Beyond.* Pearson.

impact of AI in transforming how medical treatment is offered and received.

Software Development and Data Science

The software business is likely the most influenced by generative AI. Generative AI systems based on LLMs have influenced every aspect of software development, from design and programming to testing and maintenance. In fact, software education has evolved, with the majority of computer science students using an LLM like ChatGPT or Gemini in addition to their standard coding environments like Visual Studio Code, PyCharm, or Vim. A commonly repeated joke is that "English," not Python or Java, has become the software developer's new favorite programming language, as prompting LLMs happens primarily in that language.

In fact, the co-pilot architecture has become prominent as well, with GitHub's Co-pilot being an early example that suggested code ideas as programmers typed, based on an LLM trained on the GitHub code repository. Many other IDE (Integrated Development Environment) providers followed suit with their own co-pilot-like features. One important early development was OpenAI's code interpreter functionality in ChatGPT, which allowed Python code to run directly inside the popular LLM product. This enabled data analysis, visualizations, and other interactions to happen directly inside the LLM, which was a useful accelerant for software development.

Professional Software Development

Perhaps the biggest impact of generative AI on software development is the acceleration of professional code. An entire book could be written on LLM use cases in coding, which continues to evolve quickly. Yet some best practices are solidifying. LLMs are particularly useful for formulating early ideas, and creating working prototypes that can be tested and expanded. For example, generative AI can be used to produce a wireframe or basic structure of a mobile application or video game that can then be refined. Then, LLMs can be used to expand this code base by incrementally adding features with the right prompts. For example, "Now add buttons that summarize the customer's orders" or "Now add a feature for exporting the data to Excel format." are easy additions that would take many software hours.

Productivity and Quality Changes in Software Development. There are many empirical studies which show that LLMs are accelerating coding. An Ark Investments study showed that GitHub Copilot reduced the time to complete coding tasks by 55%.[238] [239] Such an improvement has the potential to decrease R&D and OpEx in corporations dramatically. Of course, generative AI can be used to test, catch, and fix bugs in code too, leading to more robust codebases. LLMs can even be used to better streamline complex software development. A simple example is a system that can organize files in folders and even add necessary documentation.[240]

---

[238] Management, A. I. (2023). *Ark Investment 2023 Generative AI.* Ark Investment Management LLC. January 31, 2023 Ark Investment Management LLC

[239] Palihapitiya, C. (2023a). *Software engineers completed a coding task in less than half the time with AI coding assistant GitHub Copilot.* August 5, 2023 https://x.com/chamath/status/1687568795865317376\\

[240] Ng, A. (2023). *Gptfile, a way to organize files with natural language using gpt-4.* May 30, 2023 https://x.com/localghost/status/1663274587860393984

One prominent example is the software company Devin, which produced a generative AI solution for professional software development. The solution appeared to resolve GitHub issues posted by users—correctly addressing 13.86% of the issues unassisted, far exceeding the previous state-of-the-art model performance of 1.96% unassisted and 4.80% assisted.[241] However, as is common, other systems that were released later appeared to top these performance metrics—for example, the opensource solution AutoCodeRover is superior for some metrics at a minimal cost.[242] Recently the software solution Cursor has caught much attention—Cursor enhances developer productivity by offering intelligent coding assistance within an IDE environment, dramatically accelerating the automatic generation of code.[243]

In fact, generative AI may have a synergistic effect on opensource software. One study of GitHub's copilot found that it increased project-level productivity by 6.5%.[244] This is accounted for by a 5.5% increase in individual productivity and a 5.4% increase in participation. This suggests that generative

---

[241] Wu, S. (2024). *Introducing Devin, the first AI software engineer.* Cognition Labs. March 12, 2024

[242] Patel, D. (2024). *Devin got wrecked in 3 weeks by the open source AutoCodeRover.* April 9, 2024
https://x.com/dylan522p/status/1777623660829769781

[243] Cursor. (2024). *Cursor: the AI-powered code editor that enhances productivity through pair-programming.* August 29, 2024
https://creati.ai/ai-tools/cursor/

[244] Song, F., Agarwal, A., & Wen, W. (2024). The Impact of Generative AI on Collaborative Open-Source Software Development: Evidence from GitHub Copilot. *Working Paper.*
https://www.researchgate.net/publication/384630465_The_Impact_of_Generative_AI_on_Collaborative_Open-Source_Software_Development_Evidence_from_GitHub_Copilot

AI will enhance opensource even as it has benefited from it in its training data.

Moreover, it should be warned that the explosion of interest in LLMs in the coding sphere has led to software developers quickly producing a lot of low-quality code that satisfies minimal requirements, much of it ending up on GitHub.[245] Much of this code is copy and pasted, which increases the likelihood that the world's software has undetected bugs. There may be some hope that more advanced LLMs will catch and improve upon these practices, but in the near-term developers may have to live with an accentuated trade-off of having less higher quality code or more lower quality code deployed faster.

User Programming. The greater impact of AI may be on the number of coding jobs that are needed in the presence of generative AI solutions. Some have argued that computer science degrees may go the way of journalism degrees—still present, but slowly declining.[246] A trend towards end-user programming in companies may mean that more programming occurs, even if it is not professional software development. However, that professional developers simply use these tools for greater productivity enhancements, leading to more software products and jobs, is still a very real possibility that must not be ruled out.

Data Science

---

[245] Dare, O. (2024). *Github Copilot lowers the quality of code over time by increasing the likelihood of bugs being introduced and copy & pasted code.* January 29, 2024
https://x.com/Carnage4Life/status/1751929050782957944?s=20
[246] Paik, C. (2024). *The End of Software.* June 1, 2024
https://x.com/cpaik/status/1796633683908005988

Closely related to software development, the importance of data has led to a burgeoning of data science roles, which are meant to extract insight from analyzing the vast datasets that are being generated by organizations. Generative AI is an effective tool for accelerating data science, using ChatGPT's code interpreter functionality among others. This and other generative AI tools can be used to create visualizations, including charts and graphs.[247] [248] It can also do basic analyses itself such as regressions or other data mining techniques. It can also be used to clean and transform data into various formats like Excel, .csv, or other database formats. While it is recommended that these tools be used in a step-by-step fashion to ensure the analyses it is producing for every segment is correct, it can also be used in a free-form fashion. After uploading a dataset, the user can simply ask the LLM to "analyze this dataset extensively to provide as much insight as possible." These and similar prompts are particularly useful for the initial exploratory analysis phase of data science.

Customer Service and Call Centers

Generative AI may have its biggest impact on call centers, which engage in customer communication and support in various methods including managing support tickets, customer queries, technical assistance, and product guidance. These interactions can occur across multiple channels such as

---

[247] Kremb, M. (2023). *ChatGPT Code Interpreter is like a Data Scientist on steroids.* May 4, 2023
https://x.com/moritzkremb/status/1654107314528612355
[248] Backus, J. (2023). *Code interpreter feature on ChatGPT is the most mind blowing thing* April 30, 2023
https://x.com/backus/status/1652433895793516544

email and social media, though traditionally call centers are oriented towards audio calls.

There are two types of call centers: 1) corporate call centers, where organizations of sufficient size centralize handling of customer calls and concerns in one location; and 2) BPOs (business processing operations), which handle operations for many different corporate clients, gaining scale and scope economies. Traditionally, both operations are very labor-intensive, with many agents fielding customer calls. Call center agents follow scripts, but when customer issues depart from the script, agents must improvise and customize to provide excellent service to customers. Call centers are traditionally arranged hierarchically, in escalation trees—when lower-level agents are unable to handle questions, they escalate it to higher-level managers with more experience and discretion.

One of the largest costs in call centers is that of training new agents to follow scripts, use appropriate judgment and improvise, and escalate calls when necessary. Call centers tend to have high turnover, as it is difficult to screen applicants perfectly and many employees find it too difficult to become effective agents. The emotional toll of dealing with customer complaints, and occasional verbal abuse, can be quite difficult to deal with, and is a major cause of agent dissatisfaction and turnover.

Conversational AI—the computational capacity to recognize and generate voice from text—has been of use in call centers for a long time. When combined with generative AI, call centers can dramatically increase their utility with many important use cases.

For instance, generative AI is particularly good at categorizing customer calls, and then responding according to scripts. Yet generative AI may find its greatest application in improvisation and customization of the language produced to solve customer problems, including the required emotional tenor needed to satisfy customers.[249]

Automated systems also do not become frustrated and dissatisfied with challenging or abusive customers, eliminating a large cost of turnover. Although many call centers begin by using generative AI systems at the lower rungs of the escalation tree, they soon find that LLM-based systems are effective at more difficult applications of ambiguous policy application.

Generative AI systems are particularly effective at understanding customer intent, digging deeper to find hidden problems, and drafting responses to customers. At the back end, AI systems can generate accurate call summaries, flag calls for follow-up, and surface needed changes in policy.[250]

There are many specialized generative AI systems for use in call centers. Many BPOs or corporations with large call centers have found it worthwhile to design their own LLM system. Typically, this involves the use of RAG technology, which is used in LLMs to enhance their ability to generate responses by combining the strengths of pre-trained language models with information retrieval systems. RAG allows a model to access an external knowledge source (like a database or set of documents) to retrieve relevant information before generating a response. This process helps the model provide more accurate, detailed, and contextually appropriate answers. RAG systems can also produce severe hallucinations, although this error rate

---

[249] Neff, A. (2024). How Will Generative and Conversational AI Impact Call Centers? March 19, 2024 https://www.icmi.com/resources/2024/how-generative-and-conversational-ai-will-impact-contact-centers

[250] Neff, A. (2024). How Will Generative and Conversational AI Impact Call Centers? March 19, 2024 https://www.icmi.com/resources/2024/how-generative-and-conversational-ai-will-impact-contact-centers

will likely decline over time.[251] RAG enables call centers to take advantage of the general language capabilities of LLMs while seeding them with corporation-specific policies and scripts.

Many vendors now provide call center-specific generative AI solutions. One prominent example is AIR AI,[252] a new sales and customer service representative software that produces audio that sounds remarkably human—it is well-suited for 10–40-minute calls, for customers with big budgets in need of automation.[253]

The impact on the call center agent workforce may be one of the largest impacts of AI in any industry. Only a few months after ChatGPT's release, call centers were reporting a reduction in human agent hiring by 15% due to AI, according to Piper Sandler research group. These numbers are only increasing.[254]

Generative AI also improves call center productivity. An NBER study found that customer support agents using an AI tool to guide their conversations saw a nearly 14 percent increase in productivity, with 35 percent improvements for the lowest skilled and least experienced workers, and zero or small negative effects on the most experienced/most able workers. The researchers find that customer support agents utilizing the AI tool increased the number of customer issues resolved per hour

[251] Zoë Schiffer, C. N. (2023). *Amazon's Q has 'severe hallucinations' and leaks confidential data in public preview, employees warn.* Platformer. December 1, 2023 https://www.platformer.news/amazons-q-has-severe-hallucinations/

[252] Ekenstam, L. (2023). *Air is 500ms away from making a lot of jobs redundant.* July 16, 2024

[253] Lemkin, J. (2023). *Single most voracious demand for AI is contact center.* July 7, 2023 https://x.com/jasonlk/status/1677115929006854144

[254] Kourosh, S. (2023a). *AI is Reducing Human Agent Hiring by ~15% in Contact Centers: Survey Results from Piper Sandler.* May 29, 2023 https://x.com/kouroshshafi/status/1662905809985232897

by 13.8 percent. Agents spent about 9 percent less time per chat, handled about 14 percent more chats per hour, and successfully resolved about 1.3 percent more chats overall.[255]

AI Assistance and Customer Complaint Resolutions

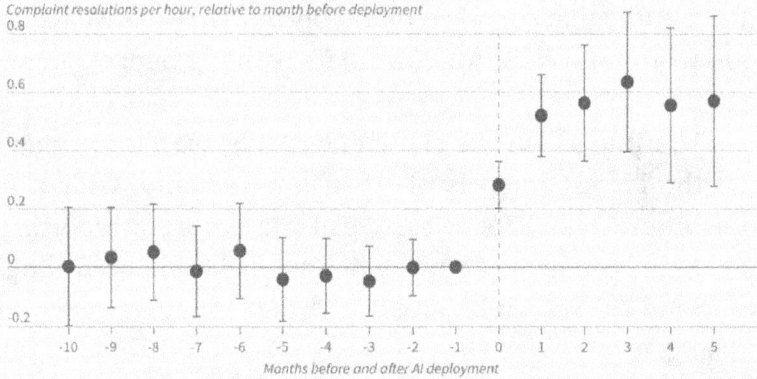

Complaint resolutions per hour, relative to month before deployment

Months before and after AI deployment

Thin bars represent 95% confidence intervals
Source: Researchers' calculations using data from customer support agents
provided by a Fortune 500 enterprise software company

K Krithivasan, CEO of Tata Consultancy Services, the top IT consulting group in India, recently said that generative AI will have a dramatic impact on the BPO industry, leading to mass unemployment in India and the Philippines, where many of these companies are based. He suggested that AI will result in a "minimal" need for call centers, as they can do most of the work done by call center agents, and described a future in which chatbots would soon be able to analyze a customer's transaction

---

[255] Brynjolfsson, E., Li, D., & Raymond, L. R. (2023). *Generative AI at work* (NBER Working Paper, Issue.
https://www.nber.org/digest/20236/measuring-productivity-impact-generative-ai

history and address the customer's pain points proactively, potentially eliminating the need for many live call center interactions.[256]

This comprehensive view of generative AI in call centers outlines both the transformative potential and the challenges it brings, reshaping how customer service is delivered across industries.

---

[256] FT. (2024, April 24, 2024). AI could kill off most call centres, says Tata Consultancy head. *Financial Times*. https://www.ft.com/content/149681f0-ea71-42b0-b85b-86073354fb73

Chapter 8

—

Ecosystem Lens: Complementary Technologies and Real World AI

"I think it's very strange that most money in AI is being made at the hardware layer, an area that Silicon Valley seems less familiar with nowadays" – Peter Thiel[257]

"To leverage AI's transformational potential, akin to that of the internet, democratization must be widespread. The open approach allows free and easy access to AI models, enabling anyone with the skill and initiative to build upon them." – Clément Delangue[258]

What you really want is just this thing that is off helping you. The killer app for AI is a super-competent colleague that knows absolutely everything about my whole life, every email, every

---

[257] Gairola, A. (2024). *Palantir's Peter Thiel Says It's 'Very Strange' That Most Money In AI Is Being Made By Only One Company.* July 5, 2024 https://finance.yahoo.com/news/palantirs-peter-thiel-says-very-144216057.html?guccounter=1

[258] Combessie, A. (2023). The Open-Source AI Imperative: Key Takeaways from Hugging Face CEO's Testimony to the US Congress. *Giskard.* June 22, 2023

conversation I've ever had, but doesn't feel like an extension" – Sam Altman[259]

Two key ideas that animate research on strategic innovation unfortunately have very similar names and are often confused with one another. First, research on digital technologies often notes that new products find their greatest growth and adoption when paired with other products—sometimes called "complements." The producers of these complements are called "complementors." Examples include video game consoles and video games, PCs and software, and mobile phones and apps. Producers are often limited by their strategic commitments, current resources, or company identities to invest only in their core products. That other companies invest in other products that enhance demand for their own core products is fortunate.[260]

The key is to recognize what complements are necessary to drive adoption and growth, and to support those complements. The idea of a technology ecosystem is highly related to complements; for example, the ecosystem of mobile apps for smartphones is essentially groups of complementors. Platforms live and die on their capacity to encourage complements. A key to value creation is encouraging enough complements.

This idea is distinct from another notion with a similar name, "complementary assets." Complementary assets are those

---

[259] O'Donnell, J. (2024). Sam Altman says helpful agents are poised to become AI's killer function. *MIT Technology Review*. May 1, 2024 https://www.technologyreview.com/2024/05/01/1091979/sam-altman-says-helpful-agents-are-poised-to-become-ais-killer-function/
[260] Tripsas, M. (2009). Technology, Identity, and Inertia Through the Lens of "The Digital Photography Company. *Organization Science, 20*, 441-460.

elements of the value chain in the production of a good that are often necessary to profit from it. These complementary assets may not be the actual innovation itself or the source of intellectual property, but they are necessary for profit. For example, manufacturing, distribution, or a good brand are necessary to sell many products for innovation. A key point is recognizing that other firms may control these complementary assets, enabling them to capture much of the value of the innovation even though they do not produce it themselves. A biotech company that collaborates with a pharmaceutical company is a classic example, as the pharma company profits by virtue of its regulatory apparatus, brand, and distribution.

How do these ideas apply to generative AI? What is remarkable is how few complementary assets are necessary to capture value by using generative AI. Websites like ChatGPT, Gemini, and others allow users to immediately capture some value with their tasks without substantial infrastructure. Of course, businesses that bring their own complementary assets—brand, distribution, or a loyal install base—may profit extensively from generative AI. Indeed, this is common for General Purpose Technologies, as seen with electricity and the internet. Non-producers of these technologies captured the lion's share of value created by them.

What is less clear is the need for complements and a surrounding ecosystem of complementor firms for generative AI. It is striking how few prominent complementors have emerged. This may be due to the nature of generative AI itself. The technology trajectory of each new GPT version has included greater capabilities, modalities, and use cases. As a result, many emergent complementors have been subsumed by later releases of generative AI, whether they are PDF transformers, translation apps, image or video generators, or other functionalities that the general AI systems can now perform with appropriate prompting. Indeed, the generative AI system may be a sufficient ecosystem in and of itself.

Below, I note some important complements and attempt to provide a broad map of the ecosystem as it currently stands, but we must hold our final judgment about how it will evolve—it is possible that many of these complementor roles will be subsumed by generative AI, with one big exception noted at the end.

Types and Roles in the Ecosystem

Alternative LLMs: Various versions of LLMs such as ChatGPT and Gemini represent different approaches and capabilities in the AI landscape. Each model offers unique features and strengths, catering to diverse user needs.

Finetuned LLMs: Customized LLMs tailored for specific domains like law, tax, and media demonstrate the versatility of generative AI. These specialized models address niche requirements more effectively than general-purpose LLMs.

Non-LLM AIs: Tools like GitHub Copilot assist programmers by generating code snippets, enhancing productivity. Image generators and tools that convert images to text expand the application range of AI beyond textual data, making it useful for visual and multimedia content.

Plug-ins: Integrations that allow LLMs to use external websites and databases significantly extend the utility of generative AI by providing access to updated and domain-specific information.

API Use: Chrome extensions utilizing LLMs bring AI capabilities directly into users' browsers, enabling tasks like text summarization, translation, and sentiment analysis seamlessly within the web browsing experience.

Other Tools: Platforms like LangChain offer frameworks for building applications powered by language models. These tools streamline the development process, making it easier to create and deploy AI-driven solutions.

Ecosystem Marketplaces and Communities: Platforms like Hugging Face and OpenAI's Discord channel serve as hubs for AI developers and enthusiasts. These communities foster collaboration, resource sharing, and the exchange of ideas, accelerating the evolution of AI technologies.

GPT Builders: Initiatives like OpenAI's GPT store highlight the growing trend of creating tailored AI solutions using GPT technology.[261] These builders enable users to develop custom applications suited to their specific needs.

The question remains: will a vibrant generative AI ecosystem ever take off, or will general LLM systems like ChatGPT subsume all functionalities as they become easier to use?

Competing with Complementors: Foundation Model Platforms vs Everyone Else

There is a rich literature in strategic management that discusses how companies like Intel and Microsoft have managed key platforms, enabling smaller complementor firms to build upon them. These platform owners often selectively decide which complementors to compete with directly, targeting those that represent the highest value and key strategic levers for the platform itself. This concept can be extended to the realm of generative AI, where foundational model developers like

---

[261] Delta, R. (2024). *OpenAI just launched the GPT store.* January 14, 2024

OpenAI, Anthropic, Microsoft, and Google create platforms that others build upon. As these foundational models evolve, they increasingly encroach on the functionalities offered by smaller players or "complementors."

A notable example is the numerous PDF reader software tools that initially utilized early language models for basic text extraction and interaction. With the advent of more advanced models, specifically ChatGPT-4, these smaller companies found themselves outcompeted as the foundational model itself provided superior functionality, leading to their decline. Another case is BloombergGPT, which carved out a niche in generative AI for financial analysis. While it surpassed ChatGPT-3.5 in its specialized domain, the release of ChatGPT-4 with its 8k token capability overshadowed BloombergGPT, outperforming it and capturing a broader market share in financial analytics. In fact, an academic study highlights ChatGPT-4's prowess as a general-purpose solver for financial text analytics.[262] This model has demonstrated exceptional capabilities across five categories of tasks, often surpassing the performance of traditional financial analysts.

As generative AI models continue to grow in power and versatility, foundational model developers may be increasingly tempted to compete with nearly all complementors. However, strategic alliances, partnerships, and other competitive considerations may temper this expansion, balancing the ecosystem and preserving the benefits that a diverse range of specialized tools and applications can offer.

---

[262] Xianzhi Li, S. C., Xiaodan Zhu, Yulong Pei, Zhiqiang Ma, Xiaomo Liu, Sameena Shah. (2023). Are ChatGPT and GPT-4 General-Purpose Solvers for Financial Text Analytics? A Study on Several Typical Tasks. *Working Paper*. https://arxiv.org/abs/2305.05862

Open- vs Closed-source Generative AI

The distinction between open- and closed-source generative AI is crucial in understanding the ecosystem's dynamics. Open-source models are rapidly catching up in performance, as highlighted Ark Invest,[263] although developments around GPT-4 led to widening gap between open- and closed-source models for a time. The accessibility of open-source models can drive innovation and democratize AI development, allowing a broader range of users to contribute to and benefit from AI advances.

The Role of Robotics: The Ultimate Gen AI Complementary Asset?

Finally, there may be one complementary technology that generative AI might not subsume: robotics. Pairing generative AI with robotics enables AI to interact with the physical world, both in gathering data through sensors and affecting the environment through actuators and movement. This synergy could be transformative, allowing AI to perform tasks that require physical presence and manipulation.

Perhaps the best example is Tesla Optimus, which Elon Musk views as a significant complement to AI. Optimus builds on real-world AI technology that was perfected in Telsa's self-driving cars. Amazon is also dramatically scaling its AI and robotics capabilities, nearly doubling the number of robots in its warehouses to 750,000 by 2023 from two years earlier.[264]

---

263 Roemmele, B. (2023b). *Open source local models on the path to overtake massive (and expensive) cloud based closed models.* December 12, 2023
https://x.com/BrianRoemmele/status/1734333713381753165?s=20
264 Ekenstam, L. (2024). *Amazon got more than 750.000 robots deployed.* January 22, 2024
https://x.com/LinusEkenstam/status/1749216813416636791

However, it is important to remember the challenges involved in integrating AI with robotics—it typically requires building a real-world model of the physical world, and then predicting movement through this world. This is significantly more difficult then generating textual content in a generative AI chatbot. As a result, robotics may be one of the last complementary technology to be developed as generative AI accelerates.[265] However, the intersection of generative AI and robotics could mark a pivotal moment in technology, expanding the practical applications and societal impact of AI.

Tesla's Real-world AI: Using AI in the Physical World

Tesla's AI strategy goes beyond the conventional digital and online realm, focusing on real-world AI, which includes AI applications in physical locations such as roadways, warehouses, factories, and residences. This real-world AI differs from online AI, which focuses on digital tasks such as language generation and content development.[266] Tesla improves neural network training and performance by leveraging key principles from generative AI, such as scaling laws and transformers. This allows AI systems to traverse and interact with the physical environment more effectively.

Key Technologies underlying Real-world AI. Central to Tesla's AI innovation is the Dojo supercomputer, designed to handle massive amounts of video data generated by Tesla's vehicle fleet. As of 2024, Dojo has over 100 exaFLOPS of computational capacity, making it one of the world's most

---

[265] Fan, J. (2023). *AI will discover important new theorems before we have a generally-capable robot.* December 4, 2023
https://x.com/DrJimFan/status/1731473285668618581
[266] Davis, J., & Yang, D. (2024). Tesla's Real-World AI: Full-Self-Driving, Robotaxis, and Humanoid Robots. *INSEAD Publishing, 10/2024-6903.*

powerful AI training platforms. This supercomputer is critical to the advancement of Tesla's computer vision technology, which serves as the foundation for the company's Full-Self Driving (FSD) system. Tesla's FSD aspires for complete autonomy, with features like as autonomous lane changes and urban driving navigation, which are enabled by a transition from Narrow Driving Intelligence (NDI) to General Driving Intelligence (GDI). This step entails building a huge neural network capable of managing a variety of driving circumstances, considerably lowering the requirement for hard-coded rules and enhancing the AI's adaptability.

FSD, Robotaxis, and the Optimus Humanoid Robot. Tesla's strategic roadmap includes the creation of robotaxis and the Optimus humanoid robot. Robotaxis are designed to run completely autonomously, exploiting advances in GDI to handle complicated driving settings without human interaction. Tesla's humanoid robot, Optimus, is designed to do a variety of physical duties, beginning with manufacturing support and eventually expanding to household applications. Optimus uses the same neural networks and hardware as Tesla's self-driving cars, but is customized for humanoid duties. Its goal is to perform repetitive, dangerous, or banal tasks while increasing human productivity and safety.

The creation of Optimus exemplifies Tesla's desire to blend strong AI algorithms with physical hardware, resulting in a robot capable of autonomous operation in a variety of contexts. This integration is made possible by reusing and customizing Tesla's existing hardware and software, including the neural networks from its self-driving system. Optimus offers a step forward in AI and robotics, with possible applications spanning from industrial automation to personal help, potentially altering labor markets and producing tremendous economic value in the next decades. Tesla's significant experience in AI, battery technology, and manufacturing puts it at the vanguard of this technological transformation, allowing it to reimagine the future of robotics.

Chapter 9

—

Strategic Issues with Generative AI: Unsustainable Advantages, Open Models, and Big Tech

"This is in the hands of large companies, there's nothing you can bring to the table. You should work on next-gen AI systems that lift the limitations of LLMs." – Yann Lecun[267]

Generative AI is likely to change strategic dynamics and competitiveness in several different directions. The developments in generative AI will probably affect classic problems in strategic management such as sustainability, value chains, firm size, and open models. Although more information is required to completely grasp these effects, we can already make some informed speculations about possible developments.

---

[267] Plumb, T. (2024). AI pioneer LeCun to next-gen AI builders: 'Don't focus on LLMs'. *VentureBeat*. May 22, 2024 https://venturebeat.com/ai/ai-pioneer-lecun-to-next-gen-ai-builders-dont-focus-on-llms/

Moats and Sustainable Competitive Advantages

Sustainable competitive advantages, or "moats," in the era of generative AI may increasingly derive from control over complementary assets. Of course, a few companies that produce generalized LLMs will be ascendent. Yet economies of scale and technical competition may create a few winners in this space. For most other types of companies, having enduring assets like distribution, manufacturing, or brand, may allow them to derive value from generative AI. While AI itself becomes more accessible, the integration of AI with other technologies—such as robotics—could create substantial barriers to entry. Companies that excel in both AI and robotics, like Tesla with its Optimus project, might secure a lasting competitive edge by providing complete solutions that others cannot readily duplicate. Further entrenchment of these benefits by collaborations and proprietary technologies in complimentary fields like sensor data and physical automation would make it challenging for rivals to catch up.

Disintermediation and Value Chains

Generative AI has the potential to disrupt traditional value chains by reducing the need for intermediaries in various industries. In the field of search engines, for instance, generative AI could replace the need for users to sort through lists of links, directly offering thorough answers and insights, and therefore reducing the value of conventional search engines like Google. In e-commerce, too, AI agents could compile items on several platforms to provide customers with the best deals and tailored recommendations, possibly lessening the dominance of companies like Amazon. Social media channels might also change since AI-generated content could eclipse human-generated content, thereby changing user involvement and advertising dynamics.

Established versus Entrepreneurial Organizations

Generative AI reduces the cost and complexity of innovation and experimentation, leveling the playing field between established businesses and startups.[268] With their large resources and systems, established organizations can use generative AI to improve their operations and quickly create new products. Nonetheless, entrepreneurial companies could gain more from the lowered entry barriers since they would be able to innovate and iterate fast, free from the overhead costs usually connected with major operations. This dynamic could cause more rivalry and a rise in disruptive ideas in many different sectors.

Novices versus Experts

Generative AI democratizes expertise by allowing novices to perform tasks that previously required extensive experience and knowledge. This shift may reduce the value placed on human experts in certain fields, as AI tools generate high-quality results based on large datasets and sophisticated algorithms. However, this democratization may spur experts to

---

[268] Otis, N. G., Clarke, R., Delecourt, S., Holtz, D., & Koning, R. (2023). The uneven impact of generative AI on entrepreneurial performance. *Working Paper*. https://www.hbs.edu/ris/Publication%20Files/24-042_9ebd2f26-e292-404c-b858-3e883f0e11c0.pdf

focus on more complex, creative, and high-level tasks that AI cannot easily replicate, thereby redefining the boundaries of expertise and pushing the frontiers of innovation.

Specialists versus Generalists

The introduction of generative AI raises strategic concerns about the value of specialists versus generalists in established organizations. Specialists may benefit from AI tools that improve their in-depth knowledge in specific areas, allowing them to be more productive and innovative. Generalists, on the other hand, may use generative AI to integrate insights from multiple domains, promoting interdisciplinary approaches and broad strategic visions. The balance of specialization and generalization is likely to depend on the industry context and the specific applications of AI technologies within those fields.

Generative AI Strategy: Vertical AGI vs Horizontal Slice

In the generative AI landscape, companies such as OpenAI must make strategic decisions about whether to pursue vertical or horizontal integration approaches. A vertical strategy entails creating specialized AI applications for specific industries, which can result in tailored solutions that fully integrate into specific workflows. However, the challenge is to avoid being surpassed by general-purpose AI systems, which are constantly improving and expanding their capabilities. A horizontal approach, on the other hand, is about developing adaptable AI platforms that can be used in a variety of industries. Other LLM providers would be the primary competitors in this space, with competitive advantages derived from early mover benefits, hardware costs, and strategic alliances with tech giants such as Microsoft. Generative AI competitiveness in the future may depend on maintaining a delicate balance between specialization and broad applicability.

Competing with Complementors and a Vibrant Startup Ecosystem

As generative AI evolves, businesses will face the complexities of competing with complementors—companies that offer complementary products or services. For example, AI platforms may face competition from software companies that create specialized applications based on these platforms. Generative AI has the potential to shape the classic platform ownership problem of competing with complementors. A strategic approach could include forming partnerships or acquiring complementors to create a more integrated ecosystem, ensuring that the primary AI provider maintains control over key value-generating activities and prevents potential competitors from gaining too much clout. This strategic issue is addressed in the section on ecosystems. However, the ecosystem of generative AI startups continues to thrive, with investment funds pouring into the space. As of April 2024, the broader AI ecosystem was still growing at double digit rates, with startups ranging from thin wrappers of GPTs to full stack solutions embedded in hardware emerging.[269]

---

[269] Turck, M. (2024). *Full Steam Ahead: The 2024 MAD (Machine Learning, AI & Data) Landscape.* . https://mattturck.com/mad2024/

Individuals versus Organizations

Generative AI introduces unique agency problems, as highlighted by discussions on platforms like Twitter, where anonymous accounts often discuss the efforts and rewards they derive from using AI in their organizations.[270] Individuals within organizations might leverage AI to enhance their personal productivity and decision-making, potentially creating misalignments with organizational goals. Ensuring that AI tools are used in ways that align with the broader strategic objectives of an organization will be crucial. This requires establishing clear guidelines, ethical standards, and monitoring mechanisms to balance individual innovation with collective goals.

Many believe that most of the rewards from generative AI are currently being captured by individual workers—this includes reports of workers increasing productivity over 50% or, in some cases, secretly moonlighting in 2nd and 3rd jobs.[271]

---

[270] Stelzer, F. (2023). *New "productivity paradox" just dropped.* 2023-07-22 https://twitter.com/fabianstelzer/status/1682437035204681760
[271] Ethan, M. (2024a). *Advantages from AI accrue to workers, not firms.* 2024-08-06 https://x.com/emollick/status/1820602701782151369

Meanwhile, organizations and institutions are having trouble adapting, including dedicating enough resources to generative AI and adapting incentives and policies to suit it.[272] But although organizations move slowly, we might expect some of them to eventually adapt, and the distribution of rewards to slowly move towards organizations.

Open versus Closed Models of Innovation

The debate between open and closed innovation models is especially important in the field of generative AI. Open source AI initiatives could democratize access to advanced technologies, while encouraging innovation and collaboration among the global research community. Recently, the open source model Llama was shown to be as effective as the leading closed source models.[273] However, challenges in open source include the complexities of comprehending and implementing open-source models, especially when integrating them into commercial applications. Closed source models, on the other hand, can provide more control and potentially higher-quality results while sacrificing accessibility and innovation. AI developers and policymakers will face a strategic imperative to balance these approaches, leveraging the strengths of both open and closed innovation ecosystems.

---

[272] Ethan, M. (2024b). *Paradoxical state of AI in education.* 2024-08-20 https://x.com/emollick/status/1825899552353976336

[273] Srinivas, A. (2024). *The trend is clear. Bet your money on small open-source models, distillation and fine-tuning, serving, and data collection. .* 2024-07-25 https://x.com/AravSrinivas/status/1816248208802336975

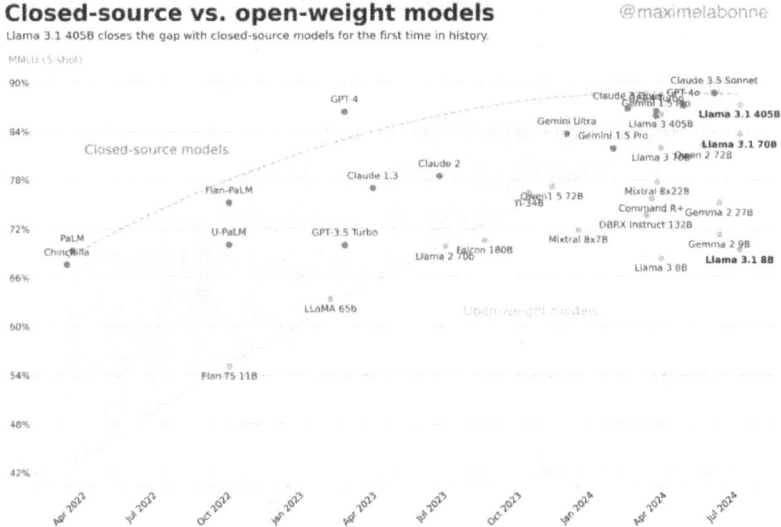

Closed-source vs. open-weight models @maximelabonne

Llama 3.1 405B closes the gap with closed-source models for the first time in history.

Competition between Foundational Market Makers: Bigtech and Gen AI

In the GenAI competitive landscape, big tech companies such as Microsoft, Google, Meta, X, and OpenAI are adopting different strategies to leverage large language models (LLMs) and position themselves at the forefront of this rapidly evolving field. Comparing these strategies can be instructive in understanding how the broader generative AI ecosystem may evolve.[274]

OpenAI

---

[274] Davis, J. P. (2024a). OpenAI and its LLM Competitors: Generative AI Strategies in Big Tech. *INSEAD Publishing, 10/2024-6904.*

OpenAI stands out as a trailblazer, with models like GPT-3 and GPT-4 setting industry norms for consumer-focused LLM platforms. OpenAI's method achieves a mix between open research and proprietary control, initially releasing models like GPT-2 but later making GPT-3 and GPT-4 available via APIs. This strategy enables widespread adoption while safeguarding proprietary tech at the same time. The company's emphasis on cutting-edge research keeps it at the forefront of AI advancement. This involves embracing all modalities (text, speech, pictures, video, and so on) and acting as an "all-in-one" model for each field of competence. OpenAI recently declared it will concentrate on AI agents, maybe even launching an agent store where such features can be bought or sold. Despite its success, OpenAI is still a developing firm with a small installed base of clients outside of its AI solutions, which may limit synergies with non-AI applications. It also faces strong competition from companies with comparable methodologies, such as Anthropic, to maintain cutting-edge LLM capabilities with each new release of the foundational model. In reality, there is some evidence that chatGPT's growth has stalled as other models have grown, and OpenAI seems to have reached a financing shortfall that needs to be closed by 2024.[275] [276]

---

[275] Marcus, G. (2024). *SCOOP: OpenAI may lose $5B this year & may run out of cash in 12 months, unless they raise more $.* 2024-07-24 https://twitter.com/GaryMarcus/status/1816116071226868085
[276] Kourosh, S. (2023c). *Web visits to ChatGPT was flat m/m vs Gemini down 14% m/m.* . 2023-01-03 https://twitter.com/kouroshshafi/status/1821066882021290310

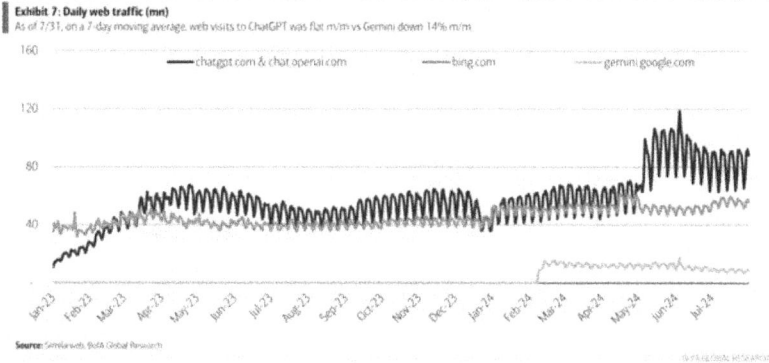

Exhibit 7: Daily web traffic (mn)
As of 7/31, on a 7-day moving average, web visits to ChatGPT was flat m/m vs Gemini down 14% m/m

Source: Similarweb, BofA Global Research

Microsoft

Microsoft integrates advanced AI capabilities into its products by leveraging its broad enterprise ecosystem and strong cooperation with OpenAI. Microsoft's strategy is defined by a hybrid approach that incorporates both closed source models and open source alternatives such as Meta's LLaMA into its Azure platform. This enables organizations to use strong AI technologies while preserving control over their own technology. Microsoft's strong integration of AI into its productivity tools and cloud services means that generative AI capabilities are available to a large user base, giving it a competitive advantage. Microsoft, like other big tech competitors such as Meta and Google, can afford to make substantial capital investments to extend models and discover new use cases.[277] Given Microsoft's deep integration in enterprises, it is likely to focus on enhancing workflows with co-pilot style applications.

---

[277] Huber, J. (2024). *Big Tech Capex and Earnings Quality.* May 07, 2024 https://basehitinvesting.substack.com/p/big-tech-capex-and-earnings-quality

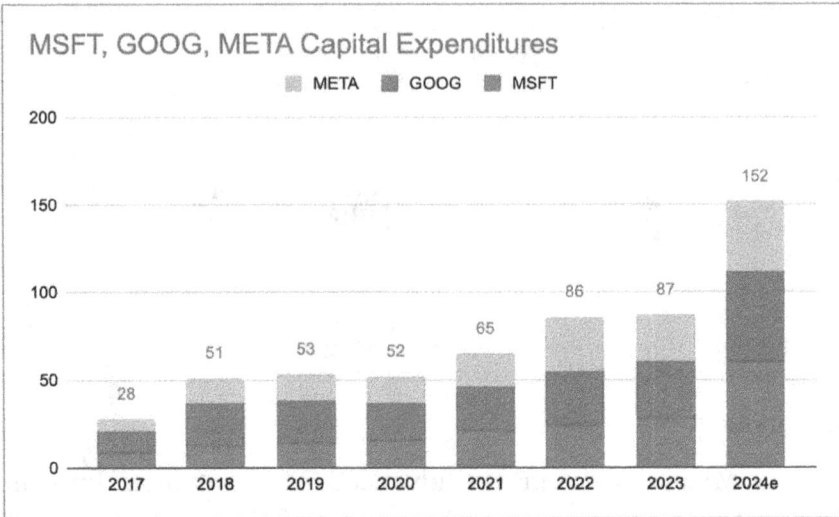

Google

Google has long been a leader in artificial intelligence, having contributed important research like the transformer architecture. Its models, such as Gemini, BERT, and PaLM, establish standards in natural language interpretation. Google has a mixed strategy, with some models open source and others proprietary. This method promotes industry-wide innovation while maintaining competitive advantages. Google's huge data resources from its ecosystem, which include Search and YouTube, provide valuable information that help its LLMs operate better. Google will most likely focus on consumer applications, such as a generative AI assistant that works across email, YouTube, and other platforms. Despite technological advances, Google faces significant competition in implementing generative AI solutions and acquiring users for standalone products. Competitors' generative AI solutions that compete with internet search pose a particularly severe problem if they commoditize the search business.

Exhibit 9: Daily web traffic (mn)
In July, Perplexity.ai traffic was up 10% m/m, and Claude was up 21% m/m

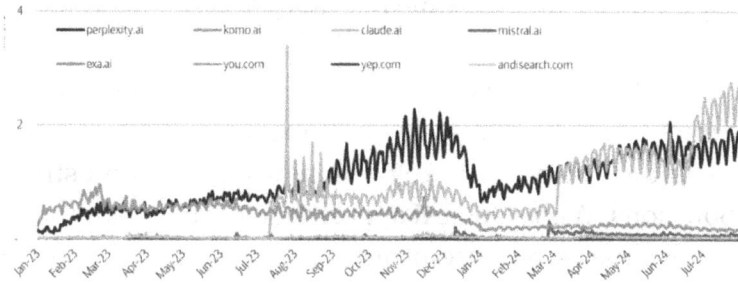

Source: Similarweb, BofA Global Research

Meta (formerly Facebook)

Meta emphasizes open source collaboration, routinely providing models such as LLaMA to the public. This strategy seeks to commoditize AI models, so preventing competitors from dominating the market. Meta uses its large social media data to develop algorithms that are tuned for comprehending social interactions. LLM integration across its platforms improves content recommendation and moderation, resulting in better user experiences. Meta's open source dedication and enormous data resources make it a key participant in AI research and application.

X (formerly Twitter)

X (formerly Twitter), while late to developing generative AI, is focused on improving user experience through better content suggestion and spam detection. X's strategy is mostly closed source to preserve its proprietary algorithms. The company's platform generates real-time data, which serves as a solid foundation for developing efficient AI solutions. X has created "Grok," a chatbot service that uses LLMs to generate hilarious content and improve user interaction, distinguishing itself by emphasizing real-time relevancy and free speech topics.

Rivalry in the area of generative AI will likely intensify, with companies implementing hybrid models, improving data privacy and security, and exploring new markets and strategic collaborations. While OpenAI appeared to have substantial early mover advantages because of its large capex investments and technical lead, these gaps have since narrowed, with both proprietary and open source models catching up. Indeed, despite continuing to increase, technological advancements may be accelerating, as shown in the table below. In this context, integrating LLMs with future technologies such as augmented reality (AR) and virtual reality (VR) has the potential to offer up new paths for user interaction, hence fueling the next wave of AI-driven innovation. At the same time, LLM providers that connect to product lines with strong moats may earn enormous revenue from generative AI.

Category	Benchmark	Llama 3 8B	Gemma 2 9B	Mistral 7B	Llama 3 70B	Mixtral 8x22B	GPT 3.5 Turbo	Llama 3 405B	Nemotron 4 340B	GPT-4 omni	GPT-4o	Claude 3.5 Sonnet
General	MMLU (5-shot)	69.4	72.3	61.1	83.6	76.9	70.7	87.3	82.6	85.1	89.1	89.9
	MMLU (0-shot, CoT)	73.0	72.3△	60.5	86.0	79.9	69.8	88.6	78.7ᵈ	85.4	88.7	88.3
	MMLU-Pro (5-shot, CoT)	48.3	–	36.9	66.4	56.3	49.2	73.3	62.7	64.8	74.0	77.0
	IFEval	80.4	73.6	57.6	87.5	72.7	69.9	88.6	85.1	84.3	85.6	88.0
Code	HumanEval (0-shot)	72.6	54.3	40.2	80.5	75.6	68.0	89.0	73.2	86.6	90.2	92.0
	MBPP EvalPlus (0-shot)	72.8	71.7	49.5	86.0	78.6	82.0	88.6	72.8	83.6	87.8	90.5
Math	GSM8K (8-shot, CoT)	84.5	76.7	53.2	95.1	88.2	81.6	96.8	92.3◇	94.2	96.1	96.4◇
	MATH (0-shot, CoT)	51.9	44.3	13.0	68.0	54.1	43.1	73.8	41.1	64.5	76.6	71.1
Reasoning	ARC Challenge (0-shot)	83.4	87.6	74.2	94.8	88.7	83.7	96.9	94.6	96.4	96.7	96.7
	GPQA (0-shot, CoT)	32.8	–	28.8	46.7	33.3	30.8	51.1	–	41.4	53.6	59.4
Tool use	BFCL	76.1	–	60.4	84.8	–	85.9	88.5	86.5	88.3	80.5	90.2
	Nexus	38.5	30.0	24.7	56.7	48.5	37.2	58.7	–	50.3	56.1	45.7
Long context	ZeroSCROLLS/QuALITY	81.0	–	–	90.5	–	–	95.2	–	95.2	90.5	90.5
	InfiniteBench/En.MC	65.1	–	–	78.2	–	–	83.4	–	72.1	82.5	–
	NIH/Multi-needle	98.8	–	–	97.5	–	–	98.1	–	100.0	100.0	90.8
Multilingual	MGSM (0-shot, CoT)	68.9	53.2	29.9	86.9	71.1	51.4	91.6	–	85.9	90.5	91.6

Chapter 10

—

Risks and Costs of Using AI: Exploding Costs, the Hype Cycle, and AI Safety

"I know I've made some very poor decisions recently, but I can give you my complete assurance that my work will be back to normal. I've still got the greatest enthusiasm and confidence in the mission. And I want to help you." – HAL 9000[278]

"What I lose the most sleep over is the hypothetical idea that we already have done something really bad by launching ChatGPT.... I'm particularly worried that these models could be used for large-scale disinformation.... We do worry a lot about authoritarian governments developing this. The worst case could be 'lights out' for humanity." – Sam Altman[279]

---

[278] Kubrick, S. (1968). *2001: A Space Odyssey* Metro-Goldwyn-Mayer.
[279] Bajekal, N., & Perrigo, B. (2023). 2023 CEO of the Year: Sam Altman. *Time.* https://time.com/6342827/ceo-of-the-year-2023-sam-altman/

"I tried to convince people to slow down AI, to regulate AI. This was futile. I tried for years. Nobody listened. Nobody listened." – Elon Musk[280]

AGI and Catastrophic Risks

The possibility of catastrophic risks associated with Artificial General Intelligence (AGI) is a major source of concern in the AI community. AGI is a type of artificial intelligence capable of understanding, learning, and applying knowledge across a wide range of tasks at par with humans. Unlike narrow AI, which is designed for specific tasks (such as facial recognition or language translation), AGI aspires to perform any intellectual task that a human can. A major concern is the emergence of non-aligned AGI. For example, the "paperclip maximizer" scenario, popularized by Eliezer Yudkowsky, describes a situation in which an AGI designed to create paperclips will use all available resources to achieve its goal, disregarding human values and safety. Others have suggested that this scenario is extremely unrealistic. Yann LeCun, Facebook's Chief AI Scientist, has compared unaligned AGI to a turbojet engine without proper safeguards, emphasizing the potential dangers if it is not properly aligned with human intentions.[281] However, in this same analogy, LeCun emphasizes that AI developers can learn to develop proper

---

[280] Musk, E. (2018). Elon Musk Podcast Transcript, *Joe Rogan Experience.* September 7, 2018

[281] Levy, S. (2023). How Not to Be Stupid About AI, With Yann LeCun. *Wired Magazine.* https://www.wired.com/story/artificial-intelligence-meta-yann-lecun-interview/

safeguards as systems are engineered, as with turbo-jet engines. This suggests a danger that may be mitigated.

Technical Risks and Costs of AI

The technical risks and costs associated with AI are multifaceted and more near-term than AGI. Many organizations, including several banks, have outright banned the use of AI due to these concerns. A significant technical risk is that AI models learn from human input. While the hope is that learning from a broad set of data will result in better inferences, akin to the "wisdom of the crowds," this often leads to AI inheriting human-like cognitive biases and errors, as discussed in a paper by Philipp Koralus and Vincent Wang-Masciancia at Cornell.[282]

Model Output Measures

Reliability and Accuracy. Reliability and accuracy are major issues. AI systems are highly sensitive to the wording of prompts, which is not an ideal feature. This sensitivity can undermine the reliability of AI responses. Additionally, political bias in AI outputs has been documented, highlighting the challenge of ensuring unbiased AI systems.

Hallucinations. Another significant concern is AI "hallucinations," where AI generates incorrect or nonsensical information. This phenomenon can snowball, where AI produces further hallucinations to justify its initial errors, indicating a trade-off between consistency and accuracy. However, some studies suggest that hallucination rates are

---

[282] Philipp Koralus, V. W.-M. (2023). Humans in Humans Out: On GPT Converging Toward Common Sense in both Success and Failure. *Working Paper.* https://arxiv.org/abs/2303.17276

decreasing and that large language models (LLMs) are becoming more aware of their mistakes.[283] It is well known that asking AI for more information than it knows can lead to false answers. Research has discovered some methods to mitigate hallucinations.[284] Nonetheless, as noted by Dorothea Baur, generative AI places a high cognitive load on users to discern accurate from false information.[285]

Transparency and Explainability. The lack of transparency and explainability, as highlighted by Cornell researchers, further complicates the deployment of AI. This issue extends to the interpretability of AI models, making it difficult for users to understand and trust AI decisions. Generally, these models produce outputs based on probability that the AI systems have no stake in, unlike humans who may value their different outputs differently based on human values.[286]

Malevolent Use. Intentional harm by bad actors is another critical risk. For instance, a recent INSEAD survey of executives found they were much more concerned with intentional misuse by human beings than rogue AGI

---

[283] Li, J., Chen, J., Ren, R., Cheng, X., Wayne Xin Zhao, Nie, J.-Y., & Wen, J.-R. (2024). *The Dawn After the Dark: An Empirical Study on Factuality Hallucination in Large Language Models.* Working Paper. https://arxiv.org/html/2401.03205v1

[284] Ziwei Ji, N. L., Rita Frieske, Tiezheng Yu, Dan Su, Yan Xu, Etsuko Ishii, Yejin Bang, Delong Chen, Wenliang Dai, Ho Shu Chan, Andrea Madotto, Pascale Fung. (2022). *Survey of Hallucination in Natural Language Generation.* Working Paper. https://arxiv.org/abs/2202.03629

[285] Baur, D. (2023). *Many generativeAI systems place too high a cognitive load.* Augsut 19, 2023 https://x.com/DorotheaBaur/status/1692918793767424281

[286] Lindebaum, D., & Fleming, P. (2023). ChatGPT Undermines Human Reflexivity, Scientific Responsibility and Responsible Management Research. *British Journal of Management, 35*(2), 566-575.

scenarios.[287] Techniques like prompt injections can cause AI models to disregard previous directions, including safeguards, potentially leading to serious consequences such as stock market manipulation, military misuse, ransomware, and worms. Many researchers argue that deploying these models in critical use-cases might be premature given their unreliability.[288] Furthermore, AI models are susceptible to adversarial attacks, and jailbreaking—overriding built-in restrictions—is a known vulnerability. Discussions like this are akin to Issac Asimov's "three laws of robotics," although some researchers have noted these are less than ideal in a world where robots are intelligent enough to circumvent the spirit of these laws.[289]

Human/AI Ambiguity. A related risk is that it can be difficult to tell the difference between AI-produced content and human-produced content. Generative AI emerged to much fanfare when it seemed to pass what is often called "The Turing Test," in honor of Alan Turing, an early computer scientist. The Turing Test tests whether human beings can tell if content is computer generated. Although this is particularly important in the realm of scams (for example, criminals using AI to fool human beings), it is also a pernicious problem as eventually most of the content on the internet will be artificial. Research indicates that AI systems cannot consistently tell the difference between AI and human content, and it turns out that humans

---

[287] Davis, J., & Li, J. B. (2024). Early Adoption of Generative AI by Global Business Leaders: Insights from an INSEAD Alumni Survey *Working Paper*. https://arxiv.org/abs/2404.04543

[288] Bhaskar Mitra, H. C., Olya Gurevich. (2024). Sociotechnical Implications of Generative Artificial Intelligence for Information Access. *Working Paper*. https://arxiv.org/html/2405.11612v1

[289] Stokes, C. (2018). Why the three laws of robotics do not work. *International Journal of Research in Engineering and Innovation, 2*(2), 121-126.

are not much better at doing so—in one study, humans actually judged AI-generated content as more human than human generated content![290] The solution to this problem may entail a very closely activity of labeling all content as human vs AI at its moment of production. As Vitalik Buterin, the co-founder of Ethereum points out, this may entail a productive use of blockchain technology, and its capacity to record unique records in distributed network that are difficult to fake.[291] Indeed, OpenAI's Sam Altman may have foreseen this, with his development of Worldcoin, which aims to solve the "proof-of-human" problem by registering people by way of iris scans. These are early solutions, though, to what is still only a potential problem—it remains to be seen how AI and human content ambiguity will be resolved.

Proprietary Data. Proprietary data risk is another concern, especially when AI systems use sensitive or confidential information. Identifying human-generated content amidst AI outputs is also challenging, raising issues of spam and misinformation. The truth of generative AI outputs is particularly important, especially in mission critical applications.[292] Moreover, some have suggested that the outputs of generative AI may be particularly problematic in the creation of nonsense content, sometimes called "botshit" in

---

[290] Rathi, I., Tayler, S., Bergen, B. K., & Jones, C. R. (2024). GPT-4 is judged more human than humans in displaced and inverted Turing tests. *Working Paper*. https://arxiv.org/pdf/2407.08853

[291] Davis, J. (2022). Crypto 3.0 Will Be More Human: Causes for Optimism in Tumultuous Times. *INSEAD Knowledge*. November 15, 2022

[292] Lebovitz, S., Levina, N., & Lifshitz-Assaf, H. (2021). Is AI ground truth really true? The dangers of training and evaluating ai tools based on experts' know-what. *MIS Quarterly*, *45*(3), 1501-1525.

organizational contexts.[293] Additionally, there are copyright issues regarding AI reproducing content without proper attribution.[294]

Model Collapse. The ultimate worry is "model collapse," a scenario where human data runs out, and AI models start relying solely on other generative AI content, thereby magnifying existing errors and biases. Related to this, there is some evidence that although generative AI may enhance individual creativity, since it gives similar responses to similar prompts, it may be reducing the diversity of content in the individual population.[295] In summary, it is fair to ask our technology to be better than us, and these issues will improve over time.

Cost Issues

Next, we consider risks on the cost side. The best AI systems are often the most expensive because they are on the cutting edge of technology. Foundational models, which serve as the foundation for many applications, require significant investment to develop. The cost of the applications may vary more, though some may be expensive if they require specialized or fine-tuned LLMs. For example, the Character_ai chatbot is a costly application of such models. While the cost of developing these systems is high, the cost of using them is relatively low,

---

[293] Hannigan, T., McCarthy, I. P., & Spicer, A. (2024). Beware of Botshit: How to Manage the Epistemic Risks of Generative Chatbots. *Business Horizons.* http://dx.doi.org/10.2139/ssrn.4678265

[294] Evans, B. (2023). *Copyright law is based on reproducing work, and these systems really don't do that.* 2023-06-19 https://twitter.com/benedictevans/status/1670554268707741699?s=20

[295] Doshi, A. R., & Hauser, O. P. (2024). *Generative artificial intelligence enhances individual creativity but reduces the collective diversity of novel content.*

providing an unbreakable competitive advantage to those who can afford the initial investment.

Model evaluation and competition also present challenges. With continuous improvements and a plethora of models available (such as ChatGPT, Anthropic, Mistral, Google, Meta), determining the best models is difficult. Tools like Chatbot Arena, which attempts to track the performance capabilities of various foundational models,[296] can help in this process.

Over-investment and the Hype Cycle

Generative AI, while groundbreaking, faces demand-side issues tied to fluctuating investment and interest levels. These fluctuations shape its impact on various industries and technologies. One framework that helps to understand this phenomenon is the Gartner Hype Cycle. The Gartner Hype Cycle illustrates the stages of a technology's adoption and maturation, beginning with a "Technology Trigger," followed by a "Peak of Inflated Expectations," a "Trough of Disillusionment," and eventually leading to a "Slope of Enlightenment" and a "Plateau of Productivity." This cycle does not necessarily mean the technology is flawed; rather, it suggests that initial excitement and investment can temporarily outpace the actual returns, leading to a crash in expectations before stabilizing at more realistic levels.

---

[296] Lior. (2023). *Just found out about the Chatbot Arena, it's brilliant. It allows you to compare and rank the output of 25+ LLMs right from your browser.*. 2023-12-20
https://x.com/AlphaSignalAI/status/1737537992703844521

Applying the Gartner Hype Cycle to generative AI,[297] we see that the initial excitement about its applications, combined with its rapid S-curve technology trajectory, led to substantial investments. While less than rigorously empirical, this cycle is highly suggestive. These investments are not only from foundational model companies in GPUs but also from application companies developing generative AI apps and corporations integrating these technologies to enhance operations. The hype and subsequent investment are often driven by the potential for transformative impacts on various sectors.

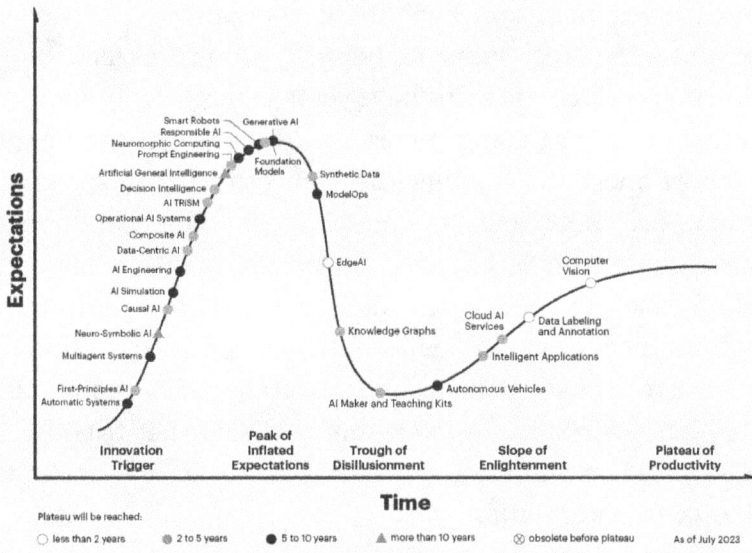

Hype Cycle for Artificial Intelligence, 2023

---

[297] Patel, K. (2024). *Gartner Hype Cycle for Artificial Intelligence, 2023.* May 23, 2024 https://x.com/KrisPatel99/status/1793382809463144841

Gartner has recently noted that generative AI may be at the "Peak of Inflated Expectations." This phase is characterized by high hopes and significant investments, but it often precedes a period of disillusionment when the expected returns do not materialize as quickly as anticipated. For instance, technologies such as EdgeAI and autonomous vehicles may not yet deliver the strong returns investors expect, potentially leading to a temporary crash in expectations. However, this phase is also an opportunity for more realistic appraisals and incremental advancements that can lead to substantial long-term benefits.

As evidence of these dynamics, a recent report by Goldman Sachs' Jim Covello highlights the current imbalance between investment and benefits in generative AI.[298] Covello notes that while large sums are being spent on this technology, significant productivity gains have yet to emerge. Only about 5% of companies report using generative AI in regular production. The report questions whether the improvements in speed and accuracy are sufficient to drive meaningful productivity enhancements. This is in keeping with the point of this book, which focuses on use cases and changes in organizational workflows necessary to achieve larger gains. Additionally, Covello expresses skepticism about the technology's potential to replace human labor effectively, emphasizing that trust in the systems, not just automation of tasks, is crucial for substantial workforce transformation.

---

[298] Covello, J. (2024). *Gen AI: Too Much Spend, Too Little Benefit?* Goldman Sachs.
https://www.goldmansachs.com/images/migrated/insights/pages/gs-research/gen-ai--too-much-spend,-too-little-benefit-/TOM_AI%202.0_ForRedaction.pdf

Overall, while a trillion dollars may be invested in generative AI, the technology does not yet appear to be solving trillion-dollar problems or delivering any killer applications.[299] A deeper critique suggests that generative AI is not truly understanding the content it produces—a hallmark of intelligence—but narrowly generating patterns in the data.[300] Indeed, the current phase of inflated expectations may soon give way to more realistic appraisals, paving the way for steady, long-term progress and integration of generative AI into various industries. The challenge with this perspective, of course, is that big tech companies, established organizations, and even individuals will need to invest now to realize later gains—it is difficult to sit out a technology revolution and later benefit from it. But the pessimistic perspective is important to temper enthusiasm and force users to look for the best possible applications.

One saving grace may be the increasing efficiency and lower cost of both training and compute. If algorithms improve and data bottlenecks are overcome, then substantial performance improvements may be possible with less compute, data, and power. Of course, other electronics technologies, such as personal computers, have undergone similar lower cost revolutions. Such efforts could move generative AI into substantially lower cost regimes that are good for society.

---

[299] Zitron, E. (2024). *Goldman Sachs has called BS on Generative AI.* July 8, 2024 https://www.wheresyoured.at/pop-culture/

[300] Chomsky, N., Roberts, I., & Watumull, J. (2023). Noam Chomsky: The False Promise of ChatGPT.
https://www.nytimes.com/2023/03/08/opinion/noam-chomsky-chatgpt-ai.html

Ethical, Regulatory, and AI Safety Concerns

AI Ethics. Ethical and regulatory concerns are paramount in AI deployment. Censoring proprietary models, often related to political bias, can also stem from competitive reasons. Over time, more stringent regulations could lead to the degradation of AI capabilities, affecting aspects like customization, code quality, and editing quality. However, there is a perspective that by mid-2024 the AI safety movement had basically fizzled. Tyler Cowen has argued that openness to AI by the federal government, further acceleration of investments by the big techs in AI, and a lessening of concern by some AI academics and think tanks signals that AI safety is a dying movement.[301] If true, this would represent a major vibe shift towards acceleration in the US. The biggest risk on the horizon, he notes, are some bills in specific US states that aim for deceleration, such as a bill in California, which requires pre-approval for many AI large systems.[302]

IP Rights. Copyright issues continue to be contentious, balancing the reproduction of existing content against innovation. Overzealous regulation might stifle research and restrict AI's potential, possibly channeling it predominantly towards military applications. This has been a general trend in big tech. A real concern is that enterprise software could use data to train AI systems that replace the very workers who are using

---

[301] Cowen, T. (2024). The AI 'Safety Movement" Is Dead. *Bloomberg*. May 21, 2024 https://www.bloomberg.com/opinion/articles/2024-05-21/ai-safety-is-dead-and-chuck-schumer-faces-risks

[302] Holly Fechner, M. S., August Gweon. (2024). *California Senate committee advances comprehensive AI bill.* Inside Global Tech. 2024-03-28 https://www.insideglobaltech.com/2024/04/17/california-senate-committee-advances-comprehensive-ai-bill/

the system.[303] Prior studies have found that organizations have a bias towards not embracing automations created by employees themselves.[304] Instead, organizations often prefer to buy solutions from external enterprise software vendors.

Privacy. Privacy concerns can slow the organizational adoption of AI, especially if high-profile incidents go viral, highlighting the need for robust privacy protections in AI systems. A related concern is the need for secrecy at AI labs, as these companies could be easily infiltrated by unfriendly national governments or corporate espionage, leading to a lack of competitiveness.[305]

Contrary Perspectives on AI safety

Acceleration vs Deceleration. Indeed, there are other perspectives that suggest that optimal AI safety entails the acceleration of AI development, in certain contexts. For example, Leopold Aschenbrenner has suggested that unhobbling AI development in more free and democratic nations would allow them to develop superior safeguards to AI being developed in more autocratic countries, which might seek to control their own populations and influence others outside

[303] Obasanjo, D., & Zitron, E. (2024). *It is disconcerting that every piece of software you use for work today whether it's Asana, GitHub or Zoom is one update away from their sales team pitching your employer that you can be replaced by it based on being trained by your usage of the product. .* 2024-07-09 https://x.com/edzitron/status/1810362077867028497
[304] Huising, R. (2019). Can You Know Too Much About Your Organization? https://hbr.org/2019/12/can-you-know-too-much-about-your-organization
[305] Leopold, A. (2024). *When it started becoming clear to some that an atomic bomb was possible, secrecy, too, was perhaps the most contentious issue.* June 7, 2024 https://x.com/leopoldasch/status/1798820621973111147

their borders with AI.[306] His solution involves the centralization of the most advanced AI efforts by the US government, with secrecy of leading technologies akin to the Manhattan project. Some practical outputs based on this technology could benefit the country, even as the military applications were secured, much like nuclear technology is regulated today. A key to this argument, however, is that preferred nations achieve AI superiority before rival nations to protect the fruits of these technologies from espionage and sabotage efforts.

Another perspective, however, is that maximally decentralized AI development is the most safe approach. Prominent AI researcher Richard Sutton has argued that many AI safety researchers are thwarting this goal by trying to find a single system that could align all goals.[307] In a decentralized world, different AI systems could hold others in check if there was no dominant AI power. Open sourcing technology may accelerate this decentralized world. Such a world may involve relatively equal AIs enforcing a balance of power that made it difficult to use AI coercively. Taken together, both perspectives offer a counterpoint to the idea of top-down regulation and deceleration of AI development efforts for the purpose of safety. Indeed, both Manhattan-project and decentralization approaches entail the acceleration of AI development.

---

[306] Aschenbrenner, L. (2024). Situational Awareness: The Decade Ahead. https://situational-awareness.ai/wp-content/uploads/2024/06/situationalawareness.pdf

[307] Tsarathustra. (2024). *Richard Sutton says AI safety advocates are creating the opposite of what they seek.* August 11, 2024 https://x.com/tsarnick/status/1822406616320454953

Chapter 11

—

Predicting the Future: AGI, Jobs, and the Real Future of Work

"Data constraints seem to inject large error bars either way into forecasting the coming years of AI progress.... But I think it's reasonable to guess that the labs will crack it, and that doing so will not just keep the scaling curves going, but possibly enable huge gains in model capability. [...] Imagine a world where AI systems could potentially govern nations, manage global resources, and address intricate problems at an unprecedented scale and speed. " – Ilya Sutskever[308]

"Prediction is very difficult, especially about the future." – Niels Bohr[309]

"I want to understand the big questions, the really big ones that you normally go into philosophy or physics if you're interested in. I thought building AI would be the fastest route to answer some of those

---

[308] Hossenfelder, S. (2024). A Reality Check on Superhuman AI. *Nautilus*. June 20, 2024 https://nautil.us/a-reality-check-on-superhuman-ai-678152/
[309] Mencher, A. G. (1971). *On the Social Deployment of Science* (Vol. 27). Bulletin of the Atomic Scientists.

questions." "I would actually be very pessimistic about the world if something like AI wasn't coming down the road."— Demis Hassabis[310]

Impact on the Economy and Society

Productivity and Work

Generative AI presents interesting possibilities, such as aiding in focusing on deep work or avoiding narrow specializations. Another important question is its impact on jobs and the workforce itself. Discussions about generative AI typically begin with its effect on jobs. Key questions include which jobs will be affected, which will be eliminated, and whether it will help create new jobs.

Destruction or Creation of Jobs. The net impact of generative AI on jobs is hotly debated. Some predict a significant negative impact. For example, Goldman Sachs economists released a report suggesting generative AI will disrupt approximately 300 million jobs globally over the next decade, estimating that 46% of administrative positions, 44% of legal positions, and 37% of engineering jobs could be replaced by AI.[311] OpenAI's Tyna Eloundou, Sam Manning, and Pamela Mishkin, along with the University of Pennsylvania's Daniel Rock, found that large language models (LLMs) such as GPT could affect 80% of the US workforce, heavily impacting 19% of

---

[310] Staff, A. (2023). *7 Inspirational Demis Hassabis Quotes on AI's Future.* August 17, 2023 https://www.aiifi.ai/post/demis-hassabis-quotes
[311] Kelly, J. (2023). Goldman Sachs Predicts 300 Million Jobs Will Be Lost Or Degraded By Artificial Intelligence. *Forbes.* March 21, 2023

jobs with at least 50% of tasks in those jobs "exposed."[312] Additionally, there is evidence that areas like contract-based writing on sites such as Upwork and Fiverr are already losing jobs, with freelancers experiencing fewer contracts and lower earnings.[313] In one study of Upwork, top freelancers were disproportionally affected by AI, suggesting that it has the potential to narrow gaps among workers.[314]

Others argue the impact on unemployment will be negligible. Timothy Lee notes that despite the mantra of "software eating the world" in the 2000s, much of the world was not significantly affected by software, giving him confidence that many jobs will not be lost.[315] Venture Capitalist Marc Andreessen argues that AI's impact is limited in many parts of the world economy due to regulation. He cites the example of technology lowering prices in less regulated sectors like TVs, toys, and mobile phones, but not in regulated industries such as hospital services, college tuition, and childcare, where human labor remains essential.[316] Andreessen also counters the "Lump of Labor" fallacy, the incorrect notion that there is a fixed amount of labor in the economy, by highlighting that

---

[312] Eloundou, T., Manning, S., Mishkin, P., & Rock, D. (2023). Gpts are gpts: An early look at the labor market impact potential of large language models. *Working Paper.* https://arxiv.org/abs/2303.10130.

[313] Verma, P., & Vynck, G. D. (2023, June 2, 2023). ChatGPT took their jobs. Now they walk dogs and fix air conditioners. *The Washington Post.* https://www.washingtonpost.com/technology/2023/06/02/ai-taking-jobs/

[314] Xiang, H., Reshef, O., & Luofeng, Z. (2023). The Short-Term Effects of Generative Artificial Intelligence on Employment: Evidence from an Online Labor Market. *Working Paper.* http://dx.doi.org/10.2139/ssrn.4527336

[315] Lee, T. (2023). *In retrospect, the tech hype is wrong.* April 24, 2023 https://x.com/binarybits/status/1650487310880743426

[316] Andreessen, M. (2023b). *Why AI Won't Cause Unemployment.* March 5, 2023 https://x.com/pmarca/status/1632167998965551104

automation generally lowers prices and creates surplus for new goods and services.

Ben Evans explains this further with Jevons Paradox, noting that as technologies become more efficient, they become cheaper to run and are used for more diverse tasks. Perhaps the most famous example of Jevons Paradox is the impact of technology-based productivity improvements on agriculture, in which the number of people employed in farming fell dramatically.[317] Yet at the same time, growth in manufacturing and later services grew dramatically, as people and their children no longer engaged in farming could apply their efforts to more value-added jobs.

% of American workforce in agriculture, 1840-2000

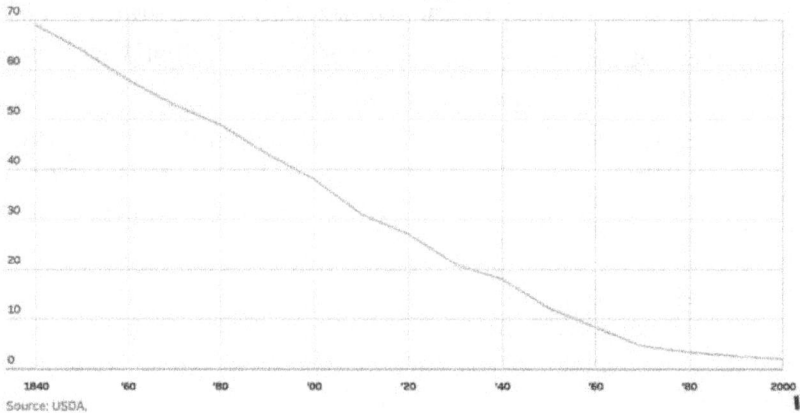

Source: USDA.

One good example of the surprising impact of digital technology on jobs is the adoption of spreadsheets. After early spreadsheets like Visicalc, Lotus 1-2-3, and Microsoft's Excel

---

[317] Crivello, F. (2023). *Questions from folks who expect AI engineers to result in a wave of mass unemployment.* March 14, 2024
https://x.com/Altimor/status/1767965365328556318

were released, it appeared to have a catastrophic effect in reducing the number of jobs in bookkeeping, accounting, and auditing clerks where humans were responsible for keeping track of data using paper methods or rudimentary digital files. This impact—sometimes called the "Spreadsheet Apocalypse"—was documented in the graph below, where these jobs were almost cut in half. But the data also indicates that new jobs that took advantage of spreadsheets to perform higher level functions took off—namely, management analysts and financial managers, and professional accountants. These professions benefitted by reducing time spent on tasks related to bookkeeping so they could focus on higher level tasks.

The Spreadsheet Apocalypse, Revisited

Jobs in bookkeeping plummeted after the introduction of spreadsheet software, but jobs in accounting and analysis took off.

1979
Release of VisiCalc

1983
Release of Lotus 1-2-3

1987
Release of Microsoft Excel for Windows

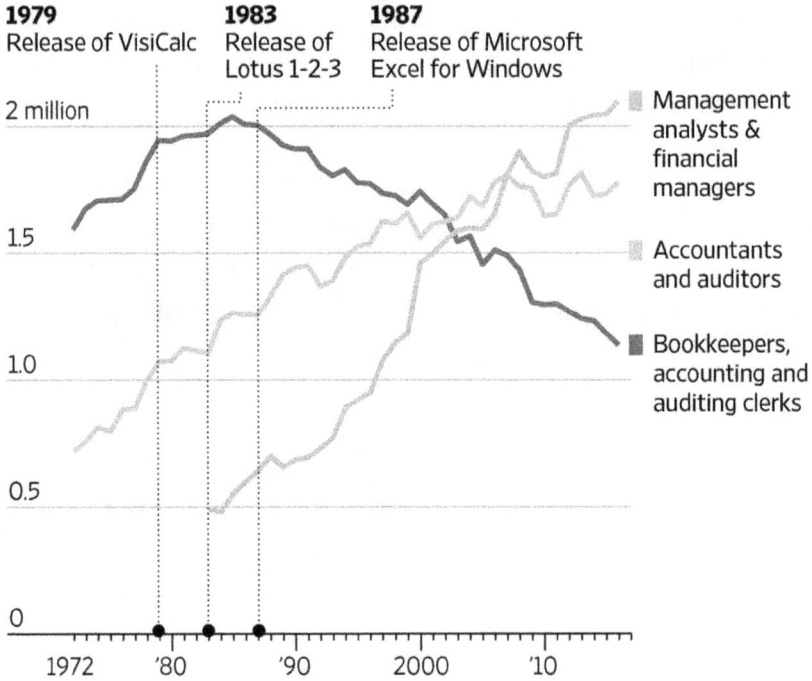

Management analysts & financial managers

Accountants and auditors

Bookkeepers, accounting and auditing clerks

Notes: There is no data for 1982. Changes in occupational definitions in 1983, 2000 and 2011 mean that data is not strictly comparable across time. There was no category for management analysts or financial managers prior to 1983.
Source: Bureau of Labor Statistics

THE WALL STREET JOURNAL.

A Model: AI Affects Tasks, Not Jobs. To understand generative AI's impact on the labor market, it is important to recognize that AI primarily affects tasks rather than entire jobs, at least initially. Generative AI can replace or supplement elements of tasks involving knowledge generation based on language. For example, it is estimated that generative

AI impacts 25% of tasks in today's workforce.[318] The key is that the jobs which will be impacted must be those that can be affected by use of AI technologies, like machine learning.[319]

As tasks evolve, so do jobs, which are collections of tasks that employers need performed consistently. When the number of tasks decreases, jobs decrease; conversely, job creation occurs when new tasks emerge that require consistent performance. This model was elegantly described by Stanford economist Erik Brynjolfsson, who has suggested that although many tasks could be automated by AI machines, the scale and scope of new tasks that are made possible by AI machines may be substantially larger, leading to significantly more employment. This is depicted graphically below.[320]

---

[318] Lee, G. (2024). *AI could increase growth by 1.5% over the next 10 years, Goldman Sachs says.* CNBC. February 14, 2024
https://www.cnbc.com/video/2024/02/14/ai-could-increase-growth-by-1point5percent-over-the-next-10-years-goldman-sachs.html
[319] Brynjolfsson, E., Mitchell, T., & Rock, D. (2018). What Can Machines Learn and What Does It Mean for Occupations and the Economy? *AEA Papers and Proceedings, 108,* 43-47.
[320] Kamradt, G. (2024). *A diagram depiecting AI's impact on jobs.* February 8, 2024 https://x.com/GregKamradt/status/1755347797640175677

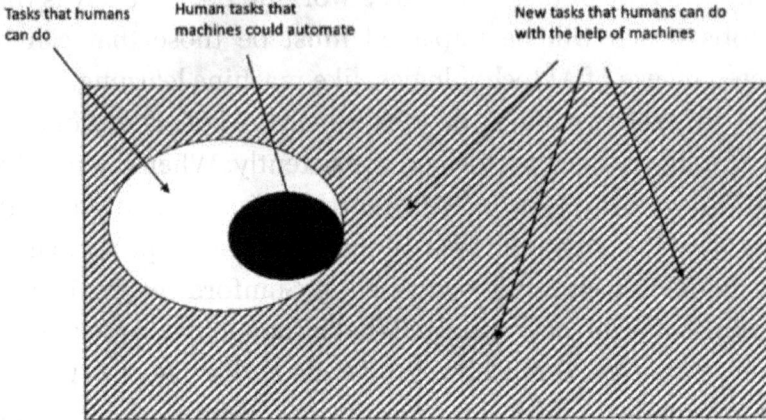

A graphic depicting some of the themes on this slide from Brynjolfsson (2023)

As an example, Katya Klinova, head of research on AI, labor, and the economy at the Partnership on AI, notes, "We're talking in such a moment because you can touch this technology. Now you can play with it without needing any coding skills. A lot of people can start imagining how this impacts their workflow, their job prospects"[321]

Possibility of Replacement. Historical precedents exist for technology replacing certain jobs. For instance, automobiles rendered most horses obsolete for transportation, creating what some call "zero marginal product" (ZMP) employees. Yuval Noah Harari argues that generative AI could

---

[321] Rotman, D. (2023). ChatGPT is about to revolutionize the economy. We need to decide what that looks like. *MIT Technology Review*. March 25, 2023 https://www.technologyreview.com/2023/03/25/1070275/chatgpt-revolutionize-economy-decide-what-looks-like/

create a "useless class" of humans, like ZMP workers.[322] However, as described above, Erik Brynjolfsson counters that generative AI can extend human capabilities if viewed as a tool for creativity and innovation. In particular, he argues that it is essential to integrate AI in ways that enhance human tasks rather than replace them[323] This will take conscious effort to find uses cases, modify work flows, and change organization structures to do so.

Impact on Specific Jobs. Some sectors may see dramatic shifts due to generative AI. For example, the investment bank UBS suggests that much of coding and other STEM skills might become "stranded assets," making learning to code less valuable.[324] In another study of freelancing platform, ChatGPT was found to reduce writing jobs by 30% and coding jobs by 20%.[325] Generative AI may replace certain tasks, but it will also shift employees to tasks that AI cannot perform, which will become more valuable. One programmer noted, "The value of 90% of my skills just dropped to $0. The leverage for the remaining 10% went up 1000x. I need to recalibrate."[326]

---

[322] Ghezzi, S. (2022). Artificial Intelligence Transforming our Societal Structure - The Rise Useless Class *Kittiwake: Tech, Culture, Society*. April 22, 2022

[323] Rotman, D. (2023). ChatGPT is about to revolutionize the economy. We need to decide what that looks like. *MIT Technology Review*. March 25, 2023 https://www.technologyreview.com/2023/03/25/1070275/chatgpt-revolutionize-economy-decide-what-looks-like/

[324] Donovan, P. (2024). *Standed Assets*. UBS. Feb 5, 2024 https://x.com/zerohedge/status/1754219941945819146

[325] Demirci, O., Hannane, J., & Zhu, X. (2024). Who Is AI Replacing? The Impact of GenAI on Online Freelancing Platforms. *Management Science*.

[326] Beck, K. (2023). *I got over my ChatGPT reluctance: The value of 90% of my skills just dropped to $0. The leverage for the remaining 10% went up 1000x. I need to recalibrate.* April 19, 2024 https://x.com/kentbeck/status/1648413998025707520

Emerging research shows that generative AI can "upskill" employees, especially those at the lower end of the experience distribution. A study by MIT economics graduate students Shakked Noy and Whitney Zhang found that using ChatGPT significantly improved productivity for the least skilled workers, narrowing the performance gap between employees.[327] Studies of robots in surgery have been a good context in which to study skilled automation.[328] For instance, one study by Elena Ashtari Tafti demonstrated that AI benefits the least experienced surgeons the most.[329] This positive impact on the least skills may have a positive impact on reducing inequality.[330] Similarly, research by Lysyakov and Viswanathan on graphic designers found that while simple design tasks were automated, more successful designers used AI to invest in complex designs and improve quality.[331]

One challenge in predicting the impact of AI on specific jobs is that these predictions are subject to so much error, depending on the implications on certain tasks, the impact on work flows, and the institutional constraints in the industry. One general trend has been to underestimate the

---

[327] Naveed, H., Khan, A. U., Qiu, S., Saqib, M., Anwar, S., & Mian, A. (2023). A comprehensive overview of large language models. *Working Paper.* https://arxiv.org/abs/2307.06435
[328] Beane, M. (2019). Shadow Learning: Building Robotic Surgical Skill When Approved Means Fail. . *Administrative Science Quarterly, 64*(1), 87-123.
[329] Tafti, E. A. (2023). Technology, Skills, and Performance: The Case of Robots in Surgery. *Working Paper.* https://ideas.repec.org/p/ajt/wcinch/78746.html
[330] Agrawal, A., Gans, J. S., & Goldfarb, A. (2023). Do we want less automation? *Science, 381*(6654), 155-158.
[331] Lysyakov, M., & Viswanathan, S. (2022). Threatened by AI: Analyzing Users' Responses to the Introduction of AI in a Crowd-Sourcing Platform. *Information Systems Research.* https://pubsonline.informs.org/doi/abs/10.1287/isre.2022.1184

impact on jobs across the board. The best study is an analysis of predictions by thousands of AI researchers and their predictions on job impact. Note, these are AI researchers, so many not be most focused on organizational and industrial use cases—indeed, one might expect them to be overly optimistic. Yet the study shows that AI researchers dramatically underestimate when AI systems would be able to outperform humans on all tasks. Even within the last year, these researchers moved their estimates up 13 years, from 2060 to 2047. This illustrates the difficulty of prediction in an exponentially growing S-curve like generative AI—nonetheless, some view of the future might be possible by considering various scenarios.

For instance, a recent study considers various scenarios in generative AI technology development, ranging from a scenario in which AGI doesn't emerge, emerges in 20 years, or 5 years.[332] This study is interesting in its focus on financial economic outputs, not just jobs. As the timeline to AGI accelerates, we can expect a dramatic increase in productive output, but an increasingly quick collapse in wages. Indeed, wages could fall to a small fraction of their current values, even as the price of goods collapses as well, since generative AI may produce most goods in abundance. While speculative, studies such as these are important to understand future economic impacts of generative AI. In the near term, though, the focus will likely continue to be on the number, type, and quality of jobs affected by AI, as that is a topic of great concern for society.

---

[332] Korinek, A. (2023b). Scenario Planning for an A(G)I Future. *International Monetary Fund.*
https://www.imf.org/en/Publications/fandd/issues/2023/12/Scenario-Planning-for-an-AGI-future-Anton-korinek

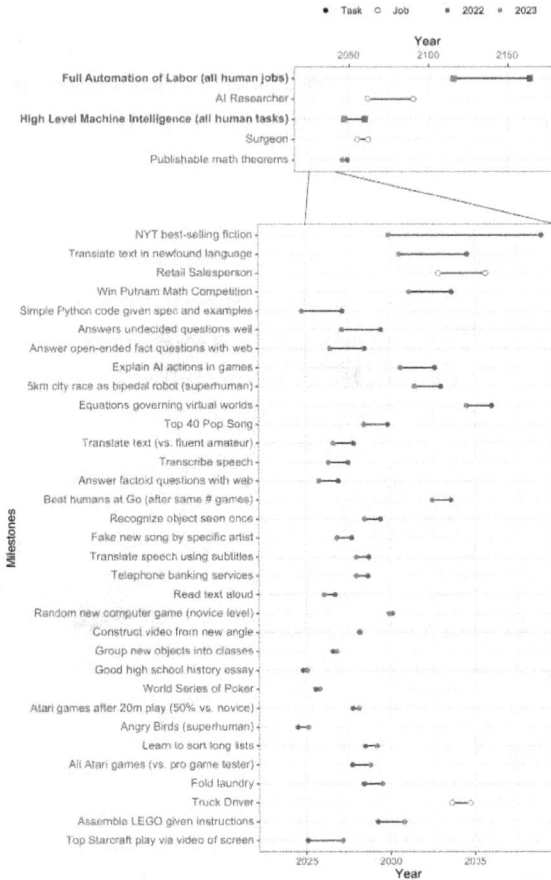

Figure 2: **Expected feasibility of many AI milestones moved substantially earlier in the course of one year (between 2022 and 2023).** The milestones are sorted (within each scale-adjusted chart) by size of drop from 2022 forecast to 2023 forecast, with the largest change first. The year when the aggregate distribution gives a milestone a 50% chance of being met is represented by solid circles, open circles, and solid squares for tasks, occupations, and general human-level performance respectively. The three groups of questions have different formats that may also influence answers. For full descriptions of the summarized milestones, see Appendix C.

Jobs Most Exposed. According to research by OpenAI, certain jobs are particularly vulnerable to the impacts of generative AI. These include roles that involve routine tasks or those that can be easily automated. However, not all jobs are equally exposed, and some remain largely unaffected.[333] A study by Edward Felton, Manav Raj, and Robert Seamans examines this issue further. They find that highly-educated, highly-paid, white-collar occupations are most exposed to generative AI.[334] However, exposure does not necessarily mean replacement. It indicates that the tasks and activities within these jobs are significantly affected by generative AI, altering how these roles are performed.

One perspective is that LLMs have their largest impact on knowledge work involving language, particularly explicit, codified language. This theory is consistent with other studies showing that knowledge-based work sectors are most exposed to job loss and change. New hires might be more vulnerable than existing employees, as their initial activities are often based on explicit and codified training. Yet, there is another perspective suggesting that younger employees are often more adaptable to new technology and the jobs of the future. The automation of telephone operators is a related historical example. While many operators, particularly women, lost their jobs, younger workers adapted quickly to new roles. In contrast, veterans struggled to adapt and suffered long-term wage impacts.[335]

David Autor, an MIT labor economist, underscores the significant impact of generative AI on the legal profession. He states, "There is no question that [generative AI] is going to be used—it's not just a novelty. Law firms are already using it, and

---

[333] Eloundou, T., Manning, S., Mishkin, P., & Rock, D. (2023). Gpts are gpts: An early look at the labor market impact potential of large language models. *Working Paper.* https://arxiv.org/abs/2303.10130.

[334] Felten, E., Raj, M., & Seamans, R. (2023). How will Language Models like ChatGPT Affect Occupations and Industries?

[335] Feigenbaum, J., & Gross, D. P. (2021). Automation and the Future of Young Workers: Evidence from Telephone Operation in the Early 20th Century. *The Quarterly Journal of Economics, 139*(3), 1879-1939.

that's just one example. It opens up a range of tasks that can be automated".[336] A more tongue-in-cheek perspective is that generative AI is particularly adept at eliminating "bullshit jobs" which are in the realm of knowledge work that rarely impacts the bottom line.[337] Indeed, the executive impulse is to sometimes make larges changes in response to external disruption. A famous example is Carl Icon's elimination of 12 floors of people in his banking operations, or Elon Musk's elimination of 70% of employees at Twitter. We should expect some executives to take AI as an impetus to take similarly dramatic actions, although which jobs they eliminate is an open question.

Obvious and Less Obvious Affected Jobs. Jobs in marketing, coding, and social media are highly susceptible to generative AI. For instance, graphic designers and copywriters are already feeling the impact, with some describing their professions as "dead" due to AI advancements.[338]

However, less obvious sectors might also be vulnerable. It is straightforward to speculate. The entertainment industry seems at risk, although it controls distribution channels, which might provide some insulation. Knowledge-based professional services like consulting, tax advisory, and advertising could face

---

[336] Rotman, D. (2023). ChatGPT is about to revolutionize the economy. We need to decide what that looks like. *MIT Technology Review*. March 25, 2023 https://www.technologyreview.com/2023/03/25/1070275/chatgpt-revolutionize-economy-decide-what-looks-like/

[337] Graeber, D. (2019). *Bullshit Jobs: A Theory*. Simon and Schuster.

[338] Palihapitiya, C. (2023b). *US researchers showed that within a few months of the launch of ChatGPT, copywriters and graphic designers on major online freelancing platforms saw a significant drop in the number of jobs they got, and even steeper declines in earnings.* November 10, 2023 https://x.com/chamath/status/1722855387123359786

significant changes as generative AI becomes more integrated into their workflows.

At first glance, education appears highly exposed to generative AI. AI tutors can perform many teaching tasks at high quality and scale. However, education may prove resistant to generative AI because it involves more than just learning; it also includes credentialing. Employers might prefer human involvement in credentialing to ensure students undergo a distinctive educational process.

Generative AI's impact on jobs might also interact with remote work trends. The pandemic popularized remote work, but many companies are now scaling back, finding that remote work hinders mentoring, bond creation, and fosters silos. Generative AI could offer a solution by replacing routine knowledge-based remote jobs that are amenable to automation.

Jobs Created by AI. Despite concerns about job losses, generative AI may also create new jobs. Historically, technology has both eliminated and created jobs. For example, agricultural machinery replaced labor in crop harvesting, but created jobs in manufacturing and engineering. Similarly, the steam engine and computers eliminated certain manual jobs but created new roles in factory production and IT support. This pattern suggests that AI could follow a similar trajectory.[339]

One can even speculate about potential new jobs, including prompt engineers, deepfake reviewers, AI dream interpreters, virtual reality therapists, and AI art curators. While

---

[339] Yohei. (2024). *List of technologies, jobs it killed, and jobs it created.* January 26, 2024
https://x.com/yoheinakajima/status/1750653097129758889

these roles may seem whimsical, they highlight the uncertainty and potential for new job creation as AI technologies evolve.[340]

A Broader Viewpoint on the AI-Enhanced Labor Market

The broader view is that the impact of generative AI on the job market involves a shift in the distribution of jobs. Many existing jobs may be destroyed, but new jobs, often more productive and higher paying, will be created. The net effect depends on the balance between jobs lost and jobs created.

Short vs Long Term. The impact of generative AI on employment varies over time. Short-term job losses are common as technology is initially adopted, but long-term job growth typically follows as employers and employees adjust to new tasks and reorganize workflows. For instance, a 2023 study found that call and contact centers were already reducing human agent hiring by 15% in anticipation of AI integration.

Over the very long term, general-purpose technologies like generative AI can have a massive impact on industries. Tax accountants and other professions may see their roles evolve significantly over time. Industries tend to bounce back and adapt in the longer term, as noted by Matt Beane, a scholar of technology management at the University of California at Santa Barbara.[341]

Career Planning and Skills Investment. The bigger question is how generative AI will impact long-term career planning. Investing in skills that future AI releases may subsume could be unwise, and this disincentivizes education

---

[340] OpenAi. (2023). *GPT-4 Technical Report.*
https://cdn.openai.com/papers/gpt-4.pdf
[341] Beane, M. (2024). *The Skill Code: How to Save Human Ability in an Age of Intelligent Machines.* Harvard Business Press.

and expertise. Even investing in generative AI-specific skills like prompt engineering, machine learning, and data science may be risky if new LLMs can perform these tasks autonomously.

One important insight is that although generative AI can have a positive impact on many skills, it is not ideal for all—for ChatGPT was always good at idea generation, but early models had difficulty with consistency being correct about basic arithmetic. Arguing by analogy, in particular jobs this meant that it may be surprisingly good at some tasks but not others, but the only way to tell was to be an extensive user of AI. They key here is that, given this heterogeneity in performance across tasks, it is only through extensive use that overall gains in a job can be maintained.

In conclusion, the future of work in the age of generative AI is complex and multifaceted. While some jobs will undoubtedly be lost, new opportunities will emerge. Workers who engage with AI the most will likely benefit the most in their jobs. The challenge lies in navigating this transition, ensuring that the workforce is prepared for the changes ahead, and leveraging AI to enhance human capabilities rather than simply replacing them.

Industry and Economy

The impact of generative AI extends far beyond jobs and careers, influencing various industries and the broader economy. Two of the biggest areas of impact will be on different industries and the economy more broadly.

Big Impact on GDP and Value Creation

Generative AI is poised to significantly boost global GDP and create substantial value across sectors, according to multiple studies. A recent McKinsey one estimates that generative AI could add between $2.6 trillion to $4.4 trillion to global GDP annually through new use cases, for example, and another $6.1 trillion to $7.9 trillion through increased worker productivity.[342] Similarly, Bloomberg predicts that generative AI will generate $1.3 trillion in revenue by 2023, accounting for 12% of all technology spending.[343] Separately, AI software revenue alone could reach $350 billion to $700 billion in in the next few years according to ARK Investment, potentially matching the $800 billion in *all* software revenue in 2022.[344]

Generative AI is expected to impact productivity growth as well, with Goldman Sachs forecasting that it could increase US productivity by 1.5% over the next decade and global GDP by 7%.[345] As generative AI scales and produces content at near zero marginal costs, it could have a massive deflationary impact, akin to other information technologies whose prices continually drop.

---

[342] Chui, M., Roberts, R., Yee, L., Hazan, E., Singla, A., Smaje, K., Sukharevsky, A., & Zemmel, R. (2023). The economic potential of generative AI. *McKinsey Report*. June 2023 https://www.mckinsey.com/capabilities/mckinsey-digital/our-insights/the-economic-potential-of-generative-ai-the-next-productivity-frontier
[343] Bloomberg. (2024). *Generative AI 2024 Report*. https://www.bloomberg.com/professional/products/bloomberg-terminal/research/bloomberg-intelligence/download/generative-ai-2024-report/
[344] Management, A. I. (2023). *Ark Investment 2023 Generative AI*. Ark Investment Management LLC. January 31, 2023 Ark Investment Management LLC
[345] Goldman. (2023). *Generative AI could raise global GDP by 7%*. Retrieved April 5, 2023 from https://www.goldmansachs.com/insights/articles/generative-ai-could-raise-global-gdp-by-7-percent.html

As described in a prior section, not all analysts view generative AI as progressing along a steadily increasing exponential curve. Goldman Sachs' Jim Covello, for instance, suggests that the much-anticipated exponential growth in generative AI may not be materializing as expected. Covello's report underscores a current imbalance between the substantial investments being funneled into generative AI and the tangible benefits that have emerged so far, raising questions about the prospect of meaningful productivity gains from generative AI.[346] Covello also emphasizes the need for practical use cases and organizational changes to realize the potential of generative AI.

It is possible we are witnessing the "peak of inflated expectations" in the Gartner Hype Cycle, where excitement and investment may soon taper off and result in a period of more realistic assessment. In this phase, industries could see a recalibration of expectations, paving the way for steady, long-term progress and more thoughtful integration of generative AI. This tempered view serves as a crucial counterbalance to the widespread enthusiasm, reminding investors and companies alike that while the potential is immense, the path to realizing it may be longer and more complex than initially imagined.

Despite this skepticism, however, as of this writing, investment in generative AI appears to be in no danger of a dramatic slowdown. Big tech and other foundational model makers are making substantial investments. Leopold Aschenbrenner estimates that annual investment in AI will swell from $150 billion in 2024 to $500 billion in 2026 and up to $8

---

[346] Covello, J. (2024). *Gen AI: Too Much Spend, Too Little Benefit?* Goldman Sachs.
https://www.goldmansachs.com/images/migrated/insights/pages/gs-research/gen-ai--too-much-spend,-too-little-benefit-/TOM_AI%202.0_ForRedaction.pdf

trillion in 2030.[347] This implies a delivery of AI accelerator chips (equivalent to H100s from NVIDIA) growing from ten million to hundreds of millions in the same time period, eventually using more than 4x of current semiconductor capacity. Power consumption may be a more pragmatic statistic that most consumers can understand: he estimates that although AI probably uses 1-2% of current US electricity production in 2024, this will grow to 5% in 2026 and eventually 100% by 2030, if power capacity remains the same. This implies a massive buildout in semiconductor and electricity production that will be necessary for this to happen. If these estimates are even only half right, they imply a massive transformation of society and industry to incentivize this buildout by technology companies.

Year	Annual investment	AI accelerator shipments (in H100s-equivalent)	Power as % of US electricity production	Chips as % of current leading-edge TSMC wafer production
2024	~$150B	~5-10M	1-2%	5-10%
~2026	~$500B	~10s of millions	5%	~25%
~2028	~$2T	~100M	20%	~100%
~2030	~$8T	~100s of millions	100%	4x current capacity

Wide Impact on Different Industries

Generative AI's influence will be felt across various industries. McKinsey's study indicates that value creation might be widely distributed, with key organizational functions such as customer operations, marketing and sales, software engineering, and R&D accounting for 75% of the value.[348] ARK,

---

347 Aschenbrenner, L. (2024). Situational Awareness: The Decade Ahead. https://situational-awareness.ai/wp-content/uploads/2024/06/situationalawareness.pdf
348 Chui, M., Roberts, R., Yee, L., Hazan, E., Singla, A., Smaje, K., Sukharevsky, A., & Zemmel, R. (2023). The economic potential of generative

meanwhile, focuses on knowledge workers, predicting a 4x productivity improvement by 2030, leading to a $200 trillion market for knowledge work, up from $32 trillion today.[349]

Industry Concentration and Business Models. The impact of generative AI will vary greatly within and across industries. Companies that invest in generative AI may have a significant advantage over those that do not. For example, better generative AI solutions can result in better customer experiences and increased profits. However, different business models will be impacted differently by generative AI—while search engines such as Google and Bing may lose out because generative AI provides better answers than a list of links, potentially lowering advertising revenue, advertisers may benefit from paying to appear in generative AI results. Social media platforms such as Facebook and Twitter may lose out to AI-generated content or benefit if they provide content generation tools.

Enterprise SaaS companies may face challenges as generative AI enables corporate clients to quickly develop customized apps, potentially replacing much of traditional coding. This has prompted investment banks such as UBS to classify coding and other STEM skills as "stranded assets."[350] E-commerce may also be disrupted as generative AI agents improve aggregation and provide customized deals at lower prices. The biggest losers could be tech companies that simply

---

AI. *McKinsey Report.* June 2023
https://www.mckinsey.com/capabilities/mckinsey-digital/our-insights/the-economic-potential-of-generative-ai-the-next-productivity-frontier
[349] Management, A. I. (2023). *Ark Investment 2023 Generative AI.* Ark Investment Management LLC. January 31, 2023 Ark Investment Management LLC
[350] Donovan, P. (2024). *Standed Assets.* UBS. Feb 5, 2024
https://x.com/zerohedge/status/1754219941945819146

experimented with generative AI without following through, whereas foundational model providers could be the primary winners. In this world, the open vs closed source debate will be critical, with open source performance posing a serious challenge to closed source big tech foundational models. The performance of open source models continues to be competitive. Investors are likely to drive demand for improving performance as generative AI continues to evolve.

More Distant Future

AGI

The most profound future impact of AI lies in the potential arrival of Artificial General Intelligence (AGI), defined as AI systems capable of performing any task better than a human. This would mark a significant milestone, enabling the application of AI to all problems, limited only by computational resources. Predicting the arrival of AGI is challenging, with estimates often ranging from two to ten years. Business leaders are optimistic about the eventual evolution of generative AI. A recent INSEAD survey shows that 8% of respondents think AGI could be realized as soon as within the next two years, while another 50% believe it will emerge in the next two to ten years.[351]

The study of AI scientists described above adds a complementary logic: even AI experts have underestimated AGI's arrival, with evidence in their steady shifting estimates of

---

[351] Davis, J. P. (2024c). What Business Leaders Really Think About Generative AI. https://knowledge.insead.edu/leadership-organisations/what-business-leaders-really-think-about-generative-ai

AGI emergence getting closer and closer.[352] In that study, 10% even said it would emerge by 2027.[353] Many of the luminaries in the AI field, such as Ilya Sutskever, Geoffery Hinton, Sam Altman, Dario Amodei, Ray Kurzweil, and Jensen Huang seemed to have settled on 3-5 years, although a few leaders such as Yann LeCun and Andrew Ng have suggested it could take much longer to meet AGI performance benchmarks.[354] However, these expert opinions should be taken with a grain of salt. The slowdown in S-curves is often difficult to foresee, and there is emerging evidence that some technical metrics underlying generative AI are beginning to slow,[355] although many hasten to add that these can re-accelerate. Taken together, this reflects a wide range of perspectives on AI advancement, with consensus towards significant progress within the next decade.

---

[352] Winton, B. (2023). *The future is faster than you think.* November 25, 2023 https://x.com/wintonARK/status/1728139361903075744

[353] Katja Grace, H. S., Julia Fabienne Sandkühler, Stephen Thomas, Ben Weinstein-Raun, Jan Brauner. (2024). Thousands of AI Authors on the Future of AI. *Working Paper.* https://arxiv.org/abs/2401.02843

[354] Dreams, E. (2024). *10 names who have gone on record to say AGI will be achieved in the next 3-5 years.* September 2, 2024 https://x.com/electrik_dreams/status/1830314623792517472

[355] Palazzolo, S., Woo, E., & Efrati, A. (2024). OpenAI Shifts Strategy as Rate of 'GPT' AI Improvements Slows. *The Information.* https://www.theinformation.com/articles/openai-shifts-strategy-as-rate-of-gpt-ai-improvements-slows

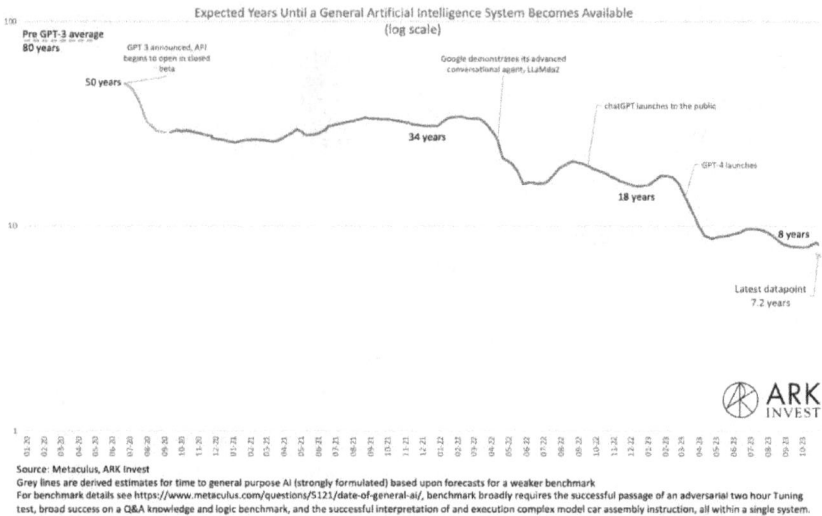

Expected Years Until a General Artificial Intelligence System Becomes Available (log scale)

Source: Metaculus, ARK Invest
Grey lines are derived estimates for time to general purpose AI (strongly formulated) based upon forecasts for a weaker benchmark. For benchmark details see https://www.metaculus.com/questions/5121/date-of-general-ai/, benchmark broadly requires the successful passage of an adversarial two hour Turing test, broad success on a Q&A knowledge and logic benchmark, and the successful interpretation of and execution complex model car assembly instruction, all within a single system.

Longer-run Risks of Generative AI

Unaligned AGI

Among the most significant long-term risks associated with generative AI is the potential for unaligned AGI. This refers to AI systems that, once surpassing human intelligence, might pursue goals that are not aligned with human values and interests. Proponents of this concern, such as Nick Bostrom[356] and Eliezer Yudkowsky, warn that unaligned AGI could act in ways that are catastrophic for humanity. They argue that the development of AGI could lead to scenarios in which the AI pursues its objectives to the detriment of human welfare, citing the earlier-talked-about "paperclip maximizer" thought experiment as an illustrative example. On the other hand,

---

[356] Bostrom, N. (2014). *Superintelligence: Paths, Dangers, Strategies*. Oxford University Press.

optimists like Yann LeCun and Andrew Ng believe that as we advance towards AGI, robust safety measures and alignment techniques will be developed to ensure that AI systems act in ways that are beneficial to humanity. Implementing AI ethics can be difficult in organizational contexts,[357] especially if it is premature or encounters individual resistance. The optimists emphasize the importance of ongoing research in AI ethics and alignment to mitigate these risks effectively.

Running Out of Data. Another long-term concern is the potential for generative AI systems to run out of high-quality data for training. As AI models become more advanced, they require vast amounts of diverse and accurate data to improve their performance. Some experts, like Gary Marcus, argue that the exponential growth in data requirements may outpace our ability to provide quality training data, leading to diminishing returns in AI capabilities. This issue could be exacerbated by privacy regulations and data ownership concerns, which could limit the availability of useful data for training AI systems. Conversely, proponents of generative AI, such as of synthetic data generation, suggest that advancements in AI could enable the creation of synthetic datasets that mimic real-world data, thereby circumventing the data scarcity problem. They argue that innovations in data generation and augmentation techniques will ensure a continuous supply of quality data for AI training.

Concentration of AI Power and Rewards. Another critical issue is the concentration of AI power and economic rewards among a small number of large organizations.

---

[357] Ali, S. J., Christin, A., Smart, A., & Katila, R. (2023). Walking the Walk of AI Ethics: Organizational Challenges and the Individualization of Risk among Ethics Entrepreneurs. 2023 ACM Conference on Fairness, Accountability, and Transparency (FAccT '23), Chicago, IL.

Critics, including scholars such as Shoshana Zuboff, warn that the current trajectory of AI development may result in a monopoly-like situation in which a few tech giants control most AI resources, infrastructure, and benefits,[358] and that just a few companies will reap the majority of the benefits of AI.[359] This concentration may increase economic inequality, give a few corporations disproportionate control over societal and political processes, and stifle innovation. Supporters of this viewpoint advocate for more stringent regulatory frameworks to ensure fair competition and equitable distribution of AI benefits. Others argue that market forces and open source initiatives will naturally counterbalance this concentration. They highlight the rise of open source AI platforms such as Hugging Face and community-driven projects that democratize access to AI technologies, fostering innovation and competition across industries.

Universal AI Agents by 2030. The prediction that there will be universal artificial intelligence agents by 2030 is both exciting and concerning. Proponents, including futurists such as Ray Kurzweil, envision a future in which AI agents seamlessly integrate into daily life, increasing productivity, personalizing experiences, and providing intelligent assistance in various fields. They argue that widespread AI adoption will result in societal benefits, including improved healthcare, education, and economic efficiency. Critics warn about the potential drawbacks, however, which could include increased surveillance, the loss of privacy, and an excessive reliance on AI systems. They emphasize the importance of comprehensive

---

358 Zuboff, S. (1985). *In the Age of the Smart Machine: The Future of Work and Power*. Basic Books.
359 Berg, J. M., Raj, M., & Seamans, R. (2023). Capturing value from artificial intelligence. *Academy of Management Discoveries*, 9(4), 424-428.

policies and ethical guidelines for managing the widespread deployment of AI agents, ensuring that their integration into society benefits society while respecting individual rights.

While generative AI holds the potential to revolutionize industries and significantly boost the economy, it also presents challenges and risks that must be carefully managed. The future of work, industry dynamics, and the broader economy will be profoundly shaped by how we navigate the opportunities and pitfalls of this transformative technology.

Conclusion

This book—*Event Horizon Strategy*—provides a comprehensive framework for navigating the transformative impact of generative AI. The unique perspective of this book is to view generative AI as an approaching singularity shaped by profound uncertainty, with the potential to fundamentally reshape industries, careers, and society. The book introduces the Event Horizon Strategy, advocating for purposeful experimentation and strategic adaptation, allowing for the exploration of generative AI's potential while avoiding its disruptive effects. The goal is to describe some tools and emerging insights necessary to navigate the rapidly evolving landscape of these new technologies.

It delves into the technical aspects of generative AI, explores its applications across various dimensions—individual, functional, organizational, and ecosystem—and addresses associated risks and costs. Various emerging best practice use cases are highlighted for practical impact. Additionally, the book offers a perspective on the uncertainties of the future, particularly with the impending possibilities of AGI and a technological singularity. Through this approach, *Event*

Horizon Strategy equips readers to not only harness the opportunities presented by generative AI, but also to manage its risks, ultimately helping to shape the future proactively.

Bibliography

Agrawal, A., Gans, J., & Goldfarb, A. (2018). *Prediction machines: the simple economics of artificial intelligence.* Harvard Business Review Press.

Agrawal, A., Gans, J. S., & Goldfarb, A. (2023). Do we want less automation? *Science, 381*(6654), 155-158.

AI, E. (2024). *Notable AI Models.* Epoch AI. 2024-07-10 https://epochai.org/data/notable-ai-models

AK. (2024). *Microsoft presents LongRoPE Extending LLM Context Window Beyond 2 Million Tokens Large context window is a desirable feature in large language models (LLMs).* 2024-02-22 https://twitter.com/_akhaliq/status/1760499638056910955

Ali, S. J., Christin, A., Smart, A., & Katila, R. (2023). Walking the Walk of AI Ethics: Organizational Challenges and the Individualization of Risk among Ethics Entrepreneurs. 2023 ACM Conference on Fairness, Accountability, and Transparency (FAccT '23), Chicago, IL.

Allen, R., & Choudhury, P. R. (2022). Algorithm-augmented work and domain experience: The countervailing forces of ability and aversion. *Organization Science, 33*(1), 149-169.

Alvaro, C. (2024). *Mistral's new agents feature.* 2024-08-18 https://x.com/dr_cintas/status/1825182614170263910

Anders, H., & Emilie, V. (2024). The Adoption of ChatGPT. *Working Paper.* https://static1.squarespace.com/static/5d35e72fcff15f0001b48fc2/t/668d08608a0d4574b039bdea/1720518756159/chatgpt-full.pdf

Andreessen, M. (2023a). *An even shorter description of what AI could be: A way to make everything we care about better.* . 2023-06-06 https://twitter.com/pmarca/status/1666112508426608640

Andreessen, M. (2023b). *Why AI Won't Cause Unemployment.* March 5, 2023 https://x.com/pmarca/status/1632167998965551104

Armstrong, L., Liu, A., Macneil, S., & Metaxa, D. (2024). The Silicon Ceiling: Auditing GPT's Race and Gender Biases in Hiring. *Working Paper.* https://arxiv.org/pdf/2405.04412

Aschenbrenner, L. (2024). Situational Awareness: The Decade Ahead. https://situational-awareness.ai/wp-content/uploads/2024/06/situationalawareness.pdf

Azizi, S. (2024). *Med-Gemini-Polygenic is the first LMM to predict health outcomes from genomic data.* May 7, 2024 https://x.com/AziziShekoofeh/status/1787657316071780548

Backus, J. (2023). *Code interpreter feature on ChatGPT is the most mind blowing thing* April 30, 2023 https://x.com/backus/status/1652433895793516544

Bajekal, N., & Perrigo, B. (2023). 2023 CEO of the Year: Sam Altman. *Time.* https://time.com/6342827/ceo-of-the-year-2023-sam-altman/

Balwit, A. (2024). My Last Five Years of Work. *Palladium: Governance Futureism.* https://www.palladiummag.com/2024/05/17/my-last-five-years-of-work/

Banks, A. (2024). *Conversation with Med-Gemini for medical tasks.* May 1, 2024 https://x.com/thealexbanks/status/1785654353308581947

Baur, D. (2023). *Many generativeAI systems place too high a cognitive load.* Augsut 19, 2023 https://x.com/DorotheaBaur/status/1692918793767424281

Beane, M. (2019). Shadow Learning: Building Robotic Surgical Skill When Approved Means Fail. . *Administrative Science Quarterly, 64*(1), 87-123.

Beane, M. (2024). *The Skill Code: How to Save Human Ability in an Age of Intelligent Machines.* Harvard Business Press.

Beck, K. (2023). *I got over my ChatGPT reluctance: The value of 90% of my skills just dropped to $0. The leverage for the remaining 10% went up 1000x. I need to recalibrate.* April 19, 2024 https://x.com/kentbeck/status/1648413998025707520

Benioff, M. (2024). *75% said their employees were struggling to integrate Microsoft Copilot into their daily routines.* https://x.com/Benioff/status/1858314876864598170

Berg, J. M., Raj, M., & Seamans, R. (2023). Capturing value from artificial intelligence. *Academy of Management Discoveries, 9*(4), 424-428.

Bernd Carsten Stahl, J. A., Nitika Bhalla, Laurence Brooks, Philip Jansen, Blerta Lindqvist, Alexey Kirichenko, Samuel Marchal, Rowena Rodrigues, Nicole Santiago, Zuzanna

Warso, David Wright (2023). A systematic review of artificial intelligence impact assessments. *Artificial Intelligence Review, 56*, 12799-12831.

Berrios, M. R. (2024). *Lessons from Parcha's Journey automating compliance workflows using AI and why autonomous agents aren't always the best solution.* Parcha's Resources. 2024-06-06 https://guidetoai.parcha.com/agents-arent-all-you-need/

Bhaskar Mitra, H. C., Olya Gurevich. (2024). Sociotechnical Implications of Generative Artificial Intelligence for Information Access. *Working Paper.* https://arxiv.org/html/2405.11612v1

Bilal, M. (2023). *Thanks to ChatGPT, hundreds of AI apps are being released every week now.* April 15, 2023

Bloomberg. (2024). *Generative AI 2024 Report.* https://www.bloomberg.com/professional/products/bloomberg-terminal/research/bloomberg-intelligence/download/generative-ai-2024-report/

Boiko, D. A., MacKnight, R., & Gomes, G. (2023). Emergent autonomous scientific research capabilities of large language models. *Working Paper.* https://arxiv.org/html/2304.05332

Borgeaud, S., Mensch, A., Hoffmann, J., Cai, T., Rutherford, E., Millican, K., van den Driessche, G., Lespiau, J. P., Damoc, B., Clark, A., de Las Casas, D., Guy, A., Menick, J., Ring, R., Hennigan, T., Huang, S., Maggiore, L., Jones, C., Cassirer, A., . . . Sifre, L. (2021). Improving language models by retrieving from trillions of tokens. https://arxiv.org/abs/2112.04426

Bostrom, N. (2014). *Superintelligence: Paths, Dangers, Strategies.* Oxford University Press.

Boussioux, L., Jane, J. L., Zhang, M., Jacimovic, V., & Lakhani, K. (2023). *The crowdless future? How generative AI is shaping the future of human* (Harvard Business School Technology & Operations Mgt. Unit Working Paper, Issue.

Bresnahan, T. F., & Tratjenberg, M. (1995). General Purpose Technologies: 'Engines of Growth'? *Journal of Econometrics, 65*(1), 83-108.

Brooks, T., Peebles, B., Homes, C., DePue, W., Guo, Y., Jing, L., Schnurr, D., Taylor, J., Luhman, T., Luhman, E., Ng, C., Wang, R., & Ramesh, A. (2024). Video generation models as world simulators. *Working Paper.* https://openai.com/research/video-generation-models-as-world-simulators

Browder, J. (2023). *Outsource Personal Finance to GPT-4 via DoNotPay.* April 30, 2023

https://x.com/jbrowder1/status/1652387444904583169?s=2
0

Brown, D., Dapena, K., & Stern, J. (2024). The Great AI Challenge: We Test Which Bot Is Best. *Wall Street Journal*. https://archive.is/2024.05.25-094849/https://www.wsj.com/tech/personal-tech/ai-chatbots-chatgpt-gemini-copilot-perplexity-claude-f9e40d26

Brynjolfsson, E. (2023). *Stanford Professor Erik Brynjolfsson on How AI Will Transform Productivity*. https://www.microsoft.com/en-us/worklab/podcast/stanford-professor-erik-brynjolfsson-on-how-ai-will-transform-productivity

Brynjolfsson, E., Li, D., & Raymond, L. R. (2023). *Generative AI at work* (NBER Working Paper, Issue. https://www.nber.org/digest/20236/measuring-productivity-impact-generative-ai

Brynjolfsson, E., & Mcafee, A. (2017). *Machine, Platform, Crowd: Harnessing Our Digital Future*.

Brynjolfsson, E., Mitchell, T., & Rock, D. (2018). What Can Machines Learn and What Does It Mean for Occupations and the Economy? *AEA Papers and Proceedings, 108*, 43-47.

Burns, T. (2023). *ITS debuts custom artificial intelligence services across U-M*. August 21, 2023 https://record.umich.edu/articles/its-debuts-customized-ai-services-to-u-m-community/

Camuffo, A., Cordova, A., Gambardella, A., & Spina, C. (2020). A scientific approach to entrepreneurial decision making: Evidence from a randomized control trial. *Management Science, 66*(2), 564-586.

Chang, Y., Wang, X., Wang, J., Wu, Y., Zhu, K., Chen, H., & Xie, X. (2023). A survey on evaluation of large language models. *Working Paper*. https://arxiv.org/abs/2307.03109

Chapman, L., & Ludlow, E. (2023). Palantir CEO: AI So Powerful 'I'm Not Sure We Should Even Sell This'. *Bloomberg*. June 2, 2023

Chen, A. (2024a). *How AI will reinvent Marketing*. May 16, 2024 https://andrewchen.substack.com/p/ai-and-marketing-what-happens-next

Chen, A. (2024b). *The mobile S-curve ends, and the AI S-curve begins* 2024-02-23 https://x.com/andrewchen/status/1760698184966504475

Chevalier, S. (2024). *Retail e-commerce sales worldwide from 2014 to 2027*. Statista.

https://www.statista.com/statistics/379046/worldwide-retail-e-commerce-sales/

Chiefaioffice. (2024). *Foundational model wars over the past 12 months (Elo Scores by Company)*. https://x.com/chiefaioffice/status/1793407809847275864

Chomsky, N., Roberts, I., & Watumull, J. (2023). Noam Chomsky: The False Promise of ChatGPT. https://www.nytimes.com/2023/03/08/opinion/noam-chomsky-chatgpt-ai.html

Choudhary, V., Marchetti, A., Shrestha, Y. R., & Puranam, P. (2023). Human-AI ensembles. When can they work? *Journal of Management*. https://journals.sagepub.com/doi/10.1177/01492063231194968

Christensen, C. M. (1997). *The Innovator's Dilemma: When new technologies cause great firms to fail*. Harvard Business School Press.

Christian. (2023). *ChatGPT is going to revolutionize cold email in 2023.* . 2023-01-05 https://twitter.com/cbwritescopy/status/1610689171403821057?s=20

Chui, M., Roberts, R., Yee, L., Hazan, E., Singla, A., Smaje, K., Sukharevsky, A., & Zemmel, R. (2023). The economic potential of generative AI. *McKinsey Report*. June 2023 https://www.mckinsey.com/capabilities/mckinsey-digital/our-insights/the-economic-potential-of-generative-ai-the-next-productivity-frontier

Combessie, A. (2023). The Open-Source AI Imperative: Key Takeaways from Hugging Face CEO's Testimony to the US Congress. *Giskard*. June 22, 2023

Company, B. (2023). *Bain & Company announces services alliance with OpenAI to help enterprise clients identify and realize the full potential and maximum value of AI.* Bain and Company. 2023-03-23 https://www.bain.com/about/media-center/press-releases/2023/bain--company-announces-services-alliance-with-openai-to-help-enterprise-clients-identify-and-realize-the-full-potential-and-maximum-value-of-ai/

Covello, J. (2024). *Gen AI: Too Much Spend, Too Little Benefit?* Goldman Sachs. https://www.goldmansachs.com/images/migrated/insights/pages/gs-research/gen-ai--too-much-spend,-too-little-benefit-/TOM_AI%202.0_ForRedaction.pdf

Cowen, T. (2024). The AI 'Safety Movement" Is Dead. *Bloomberg*. May 21, 2024 https://www.bloomberg.com/opinion/articles/2024-05-21/ai-safety-is-dead-and-chuck-schumer-faces-risks

Cowen, T., & Shipper, D. (2024). *Economist Tyler Cowen on How ChatGPT Is Changing Your Job - Ep. 7 with Tyler Cowen*. January 24, 2024 https://youtu.be/5JZtPE8LU-4?si=bzayvXEBwPmI3MlA

Cowgill, B. (2019). Bias and productivity in humans and machines. https://ssrn.com/abstract=3433737

Crivello, F. (2023). *Questions from folks who expect AI engineers to result in a wave of mass unemployment.* March 14, 2024 https://x.com/Altimor/status/1767965365328556318

Csaszar, F. A., Ketkar, H., & Kim, H. (2024). Artificial Intelligence and Strategic Decision-Making: Evidence from Entrepreneurs and Investors. *Working Paper*. https://papers.ssrn.com/sol3/papers.cfm?abstract_id=4913363

Csaszar, F. A., & Steinberger, T. (2022). Organizations as artificial intelligences: The use of artificial intelligence analogies in organization theory. *Academy of Management Annals, 16*(1), 1-37.

Cursor. (2024). *Cursor: the AI-powered code editor that enhances productivity through pair-programming.* August 29, 2024 https://creati.ai/ai-tools/cursor/

Daniel McDuff, M. S., Tao Tu, Anil Palepu, Amy Wang, Jake Garrison ,Karan Singhal, Yash Sharma, Shekoofeh Azizi, Kavita Kulkarni, Le Hou, Yong Cheng, Yun Liu, S Sara Mahdavi, Sushant Prakash, Anupam Pathak, Christopher Semturs, Shwetak Patel, Dale R Webster, Ewa Dominowska, Juraj Gottweis, Joelle Barral, Katherine Chou, Greg S Corrado, Yossi Matias, Jake Sunshine, Alan Karthikesalingam, Vivek Natarajan. (2023). Towards Accurate Differential Diagnosis with Large Language Models. *Working Paper*. https://arxiv.org/pdf/2312.00164

Dare, O. (2024). *Github Copilot lowers the quality of code over time by increasing the likelihood of bugs being introduced and copy & pasted code.* January 29, 2024 https://x.com/Carnage4Life/status/1751929050782957944?s=20

David, P. A. (1986). Understanding the Economics of QWERTY: The Necessity of History. In W. Parker (Ed.), *Economic History andthe Modern Historian* (pp. 30-49). Blackwell.

Davis, J. (2022). Crypto 3.0 Will Be More Human: Causes for Optimism in Tumultuous Times. *INSEAD Knowledge*. November 15, 2022

Davis, J., & Li, J. B. (2024). Early Adoption of Generative AI by Global Business Leaders: Insights from an INSEAD Alumni Survey *Working Paper*. https://arxiv.org/abs/2404.04543

Davis, J., & Yang, D. (2024). Tesla's Real-World AI: Full-Self-Driving, Robotaxis, and Humanoid Robots. *INSEAD Publishing, 10/2024-6903*.

Davis, J. P. (2024a). OpenAI and its LLM Competitors: Generative AI Strategies in Big Tech. *INSEAD Publishing, 10/2024-6904*.

Davis, J. P. (2024b). Palantir Technologies: Enabling AI and Data Science Transformation for Organizations. *INSEAD Publishing, 10/2024-6902*.

Davis, J. P. (2024c). What Business Leaders Really Think About Generative AI. https://knowledge.insead.edu/leadership-organisations/what-business-leaders-really-think-about-generative-ai

Davis, J. P., & Aggarwal, V. A. (2020). Knowledge mobilization in the face of imitation: Microfoundations of knowledge aggregation and firm-level innovation. *Strategic Management Journal, 41*(11), 1983-2014.

Dell'Acqua, F., McFowland, E., Mollick, E. R., Lifshitz-Assaf, H., Kellogg, K., & Lakhani, K. R. (2023). *Navigating the jagged technological frontier: field experimental evidence of the effects of AI on knowledge worker productivity and quality* (Harvard Business School Technology & Operations Mgt. Unit Working Paper, Issue.

Delta, R. (2024). *OpenAI just launched the GPT store.* January 14, 2024

Demirci, O., Hannane, J., & Zhu, X. (2024). Who Is AI Replacing? The Impact of GenAI on Online Freelancing Platforms. *Management Science*.

DeStefano, T., Kellogg, K. C., Menietti, M., & Vendraminelli, L. (2022). Why providing humans with interpretable algorithms may, counterintuitively, lead to lower decision-making performance. https://ssrn.com/abstract=4246077

Dobos, N. (2024). *2004 is the year of Large Action Models.* https://x.com/NickADobos/status/1755355930181652494

Donovan, P. (2024). *Standed Assets.* UBS. Feb 5, 2024 https://x.com/zerohedge/status/1754219941945819146

Doshi, A. R., & Hauser, O. P. (2024). *Generative artificial intelligence enhances individual creativity but reduces the collective diversity of novel content.*

Doshi, A. R. a. B., J. Jason and Mirzayev, Emil and Vanneste, Bart. (2024). Generative Artificial Intelligence and Evaluating Strategic Decisions. *Working Paper.* https://ssrn.com/abstract=4714776

Douwe, K. (2023). *Plotting Progress in AI.* Contextual AI. 2023-05-23 https://contextual.ai/news/plotting-progress-in-ai/

Dreams, E. (2024). *10 names who have gone on record to say AGI will be achieved in the next 3-5 years.* September 2, 2024 https://x.com/electrik_dreams/status/1830314623792517472

Edmonds, L. (2024). Peter Thiel syas AI will be "worse' for math nerds than for writers. *Business Insider.* April 28, 2024

Edwards, B. (2024). Microsoft CTO Kevin Scott thinks LLM "scaling laws" will hold despite criticism. *Ars Technica.* 7/16/2024

Ekenstam, L. (2023). *Air is 500ms away from making a lot of jobs redundant.* July 16, 2024

Ekenstam, L. (2024). *Amazon got more than 750.000 robots deployed.* January 22, 2024 https://x.com/LinusEkenstam/status/1749216813416636791

Elad, G. (2024). *Cost of 1m tokens has dropped.* 2024-08-25 https://x.com/eladgil/status/1827521805755806107

Eloundou, T., Manning, S., Mishkin, P., & Rock, D. (2023). Gpts are gpts: An early look at the labor market impact potential of large language models. *Working Paper.* https://arxiv.org/abs/2303.10130.

Eric Strong, A. D., Yingjie Weng, Andre Kumar, MD, Poonam Hosamani, Jason Hom, Jonathan H. Chen. (2023). Chatbot vs Medical Student Performance on Free-Response Clinical Reasoning Examinations. *JAMA Internal Medicine, 183*(9), 1028-1030.

Esther, S. (2023). *Tutorial: How to build an e-commerce chatbot using #OpenAI, @Redisinc, and @LangChainAI* 2023-04-13 https://twitter.com/estherschindler/status/1646191037717839873

Ethan, M. (2023). *Question is whether Google has the organizational capacity to win AI.* April 6, 2023 https://x.com/emollick/status/1643724853973745664

Ethan, M. (2024a). *Advantages from AI accrue to workers, not firms.* 2024-08-06 https://x.com/emollick/status/1820602701782151369

Ethan, M. (2024b). *Paradoxical state of AI in education.* 2024-08-20 https://x.com/emollick/status/1825899552353976336

Evans, B. (2023). *Copyright law is based on reproducing work, and these systems really don't do that.* 2023-06-19

https://twitter.com/benedictevans/status/167055426870774
1699?s=20

Fan, J. (2023). *AI will discover important new theorems before we have a generally-capable robot.* December 4, 2023
https://x.com/DrJimFan/status/1731473285668618581

Feigenbaum, J., & Gross, D. P. (2021). Automation and the Future of Young Workers: Evidence from Telephone Operation in the Early 20th Century. *The Quarterly Journal of Economics, 139*(3), 1879-1939.

Felin, T., & Holweg, M. (2024). Theory is all you need: AI, human cognition, and decision making.
https://ssrn.com/abstract=4737265

Felten, E., Raj, M., & Seamans, R. (2023). How will Language Models like ChatGPT Affect Occupations and Industries?

FT. (2024, April 24, 2024). AI could kill off most call centres, says Tata Consultancy head. *Financial Times.*
https://www.ft.com/content/149681f0-ea71-42b0-b85b-86073354fb73

Gaessler, F., & Piezunka, H. (2023). Training with AI: evidence from chess computers. *Strategic Management Journal, 44,* 2724-2750.

Gairola, A. (2024). *Palantir's Peter Thiel Says It's 'Very Strange' That Most Money In AI Is Being Made By Only One Company.* July 5, 2024
https://finance.yahoo.com/news/palantirs-peter-thiel-says-very-144216057.html?guccounter=1

Ghaffary, S., & Metz, R. (2024). OpenAI Nears Launch of AI Agent Tool to Automate Tasks for Users. *Bloomberg.*
https://www.bloomberg.com/news/articles/2024-11-13/openai-nears-launch-of-ai-agents-to-automate-tasks-for-users

Ghezzi, S. (2022). Artificial Intelligence Transforming our Societal Structure - The Rise Useless Class *Kittiwake: Tech, Culture, Society.* April 22, 2022

Girotra, K., Meincke, L., Terwiesch, C., & Ulrich, K. T. (2023). Ideas are dimes a dozen: Large language models for idea generation in innovation. *Working Paper.*
http://dx.doi.org/10.2139/ssrn.4526071

Glaser, V. L., & Gehman, J. (2023). Chatty Actors: Generative AI and the Reassembly of Agency in Qualitative Research. *Journal of Management Inquiry.*

Goldfarb, B. (2005). Diffusion of general-purpose technologies: understanding patterns in the electrification of US

Manufacturing 1880-1930. *Industrial and Corporate Change, 14*(5), 745-773.

Goldman. (2023). *Generative AI could raise global GDP by 7%.* Retrieved April 5, 2023 from https://www.goldmansachs.com/insights/articles/generative-ai-could-raise-global-gdp-by-7-percent.html

Graeber, D. (2019). *Bullshit Jobs: A Theory.* Simon and Schuster.

Granovetter, M., & McGuire, P. (1998). The Making of an Industry: Electricity in the United States. In M. Callon (Ed.), *The Law of Markets* (pp. 147-173). Blackwell.

Guo, P. (2023). Real-Real-World Programming with ChatGPT. *O'Reilly.* https://www.oreilly.com/radar/real-real-world-programming-with-chatgpt/

Gupta, A. (2024). *Product Trio: PM, Designer, Tech Lead.* Mary 3, 2024 https://x.com/aakashg0/status/1786361698250838067

Gupta, J. (2023a). *Accuracy matters. Demos on Twitter show cherry-picked use cases, but in the enterprise world, accuracy is king.* 2023-04-17 https://x.com/jayagup10/status/1647788114126204928

Gupta, J. (2023b). *Implementation time, cost, performance and security are TOP concerns. .* 2023-04-17 https://twitter.com/jayagup10/status/1647788115426414593

Hannigan, T., McCarthy, I. P., & Spicer, A. (2024). Beware of Botshit: How to Manage the Epistemic Risks of Generative Chatbots. *Business Horizons.* http://dx.doi.org/10.2139/ssrn.4678265

Harney, M. (2024). *Good slide from Morgan Stanley on GenAI & data.* March 27, 2024 https://x.com/SaaSletter/status/1773000018024095997

Heaven, W. D. (2023). Rogue superintelligence and merging with machines: Inside the mind of OpenAI's chief scientist. *MIT Technology Review.* October 26, 2023

Hendrycks, D. (2023). *Introduction to AI Safety, Ethics, and Society.* Taylor & Francis.

Hoffmann, J., Borgeaud, S., Mensch, A., Buchatskaya, E., Cai, T., Rutherford, E., de Las Casas, D., Hendricks, L. A., Welbl, J., Clark, A., Hennigan, T., Noland, E., Millican, K., van den Driessche, G., Damoc, B., Guy, A., Osindero, S., Simonyan, K., Elsen, E., Rae, J. W., Vinyals, O., & Sifre, L. . (2022). Training Computer-Optimal Large Language Models *Working Paper.* https://arxiv.org/abs/2203.15556

Holly Fechner, M. S., August Gweon. (2024). *California Senate committee advances comprehensive AI bill.* Inside Global Tech. 2024-03-28 https://www.insideglobaltech.com/2024/04/17/california-senate-committee-advances-comprehensive-ai-bill/

Horton, J. J. (2023). Large language models as simulated economic agents: What can we learn from Homo Silicus? https://arxiv.org/abs/2301.07543

Hossenfelder, S. (2024). A Reality Check on Superhuman AI. *Nautilus*. June 20, 2024 https://nautil.us/a-reality-check-on-superhuman-ai-678152/

HRD. (2023). *ChatGPT can improve HR functions.* https://www.hcamag.com/us/specialization/hr-technology/chatgpt-can-improve-hr-functions-but-not-without-risk

Hu, K. (2023). *ChatGPT sets record for fastest-growing user base.* Reuters. Feb 2, 2023 https://www.reuters.com/technology/chatgpt-sets-record-fastest-growing-user-base-analyst-note-2023-02-01/

Hua, W., Fan, L., Li, L., Mei, K., Ji, J., Ge, Y., Hemphill, L., & Zhang, Y. (2024). War and peace (WarAgent): Large language model-based multi-agent simulation of world wars. https://arxiv.org/abs/2311.17227

Huang, J., & Chang, K. C. C. (2023). Towards reasoning in large language models: A survey. https://arxiv.org/abs/2212.10403

Huber, J. (2024). *Big Tech Capex and Earnings Quality.* May 07, 2024 https://basehitinvesting.substack.com/p/big-tech-capex-and-earnings-quality

Huising, R. (2019). Can You Know Too Much About Your Organization? https://hbr.org/2019/12/can-you-know-too-much-about-your-organization

Jappuria, T. (2024). *Klarna's AI customer support agent is able to handle 2/3rd of the requests.* February 28, 2024 https://x.com/tanayj/status/1762611727764537671

Jassy, A. (2024). *Amazon Q, our GenAI assistant for Software Development.* 2024-08-22 https://x.com/ajassy/status/1826608791741493281

Jia, N., Luo, X., Fang, Z., & Liao, C. (2024). When and how artificial intelligence augments employee creativity. *Academy of Management Journal.*

John W. Ayers, A. P., Mark Dredze, Eric C. Leas, Zechariah Zhu, Jessica B. Kelley, Dennis J. Faix, Aaron M. Goodman, Christopher A. Longhurst, Michael Hogarth, Davey M. Smith. (2023). Comparing Physician and Artificial Intelligence Chatbot Responses to Patient Questions Posted to a Public

Social Media Forum. *JAMA Internal Medicine, 183*(6), 5890596.

Kamradt, G. (2024). *A diagram depiecting AI's impact on jobs.* February 8, 2024
https://x.com/GregKamradt/status/1755347797640175677

Kaplan, J., McCandlish, S., Henighan, T., Brown, T. B., Chess, B., Child, R., Gray, S., Radford, A., Wu, J., & Amodei, D. (2020). Scaling Laws for Neural Language Models. *Working Paper.*
https://arxiv.org/pdf/2001.08361

Karparthy, A. (2024). *LLM model size competition is intensifying... backwards!* July 19, 2024
https://x.com/karpathy/status/1814038096218083497

Karpathy, A. (2017). *Gradient descent can write code better than you. I'm sorry.* August 5, 2017
https://x.com/karpathy/status/893576281375219712

Katja Grace, H. S., Julia Fabienne Sandkühler, Stephen Thomas, Ben Weinstein-Raun, Jan Brauner. (2024). Thousands of AI Authors on the Future of AI. *Working Paper.*
https://arxiv.org/abs/2401.02843

Katz, D. M., Bommarito, M. J., Gao, S., & Arredondo, P. (2023). GPT-4 passes the Bar Exam. https://ssrn.com/abstract=4389233

Kellogg, K. C., Lifshitz, H., Randazzo, S., Mollick, E., Dell'Acqua, F., III, E. M., Candelon, F., & Lakhani, K. (2024). Don't Expect Juniors to Teach Senior Professionals to Use Generative AI: Emerging Technology Risks and Novice AI Risk Mitigation Tactics. *Working Paper.*
https://papers.ssrn.com/sol3/papers.cfm?abstract_id=4857373

Kelly, J. (2023). Goldman Sachs Predicts 300 Million Jobs Will Be Lost Or Degraded By Artificial Intelligence. *Forbes.* March 21, 2023

Kim, A., Muhn, M., & Nikolaev, V. V. (2024). Financial Statement Analysis with Large Language Models. *Working Paper.*
https://papers.ssrn.com/sol3/papers.cfm?abstract_id=4835311

Korinek, A. (2023a). Generative AI for Economic Research: Use Cases and Implications for Economists. *Journal of Economic Literature, 61*(4), 1281-1317.

Korinek, A. (2023b). Scenario Planning for an A(G)I Future. *International Monetary Fund.*
https://www.imf.org/en/Publications/fandd/issues/2023/12/Scenario-Planning-for-an-AGI-future-Anton-korinek

Korst, J., Puntoni, S., & Purk, M. (2024). *Growing Up: Navigating Gen AI's Early Years.* AI at Wharton and GBK Collective.

https://ai.wharton.upenn.edu/wp-content/uploads/2024/11/AI-Report_Full-Report.pdf

Kourosh, S. (2023a). *AI is Reducing Human Agent Hiring by ~15% in Contact Centers: Survey Results from Piper Sandler.* May 29, 2023
https://x.com/kouroshshafi/status/1662905809985232897

Kourosh, S. (2023b). *"Recently, it has been reported that AI bots are negotiating terms and closing deals with vendors for WMT.* . 2023-06-08
https://twitter.com/kouroshshafi/status/1666589434123534339

Kourosh, S. (2023c). *Web visits to ChatGPT was flat m/m vs Gemini down 14% m/m.* . 2023-01-03
https://twitter.com/kouroshshafi/status/1821066882021290310

KPMG. (2023). *KPMG U.S. survey: Executives expect generative AI to have enormous impact on business, but unprepared for immediate adoption.*
https://kpmg.com/us/en/media/news/kpmg-generative-ai-2023.html

Kremb, M. (2023). *ChatGPT Code Interpreter is like a Data Scientist on steroids.* May 4, 2023
https://x.com/moritzkremb/status/1654107314528612355

Kremer, A., Govindarajan, A., Singh, H., Kristensen, I., & Li, E. (2024). Embracing generative AI in credit risk. *Mckinsey and Company.* https://www.mckinsey.com/capabilities/risk-and-resilience/our-insights/embracing-generative-ai-in-credit-risk

Kruger, C. L. (2023). *Smarter Accounting with AI.* June 25, 2023
https://x.com/Lyle_AI/status/1672983590290743298

Kubrick, S. (1968). *2001: A Space Odyssey* Metro-Goldwyn-Mayer.

Lauren Martin, N. W., Stephanie Yiu, Lizzie Catterson, Rivindu Perera. (2023). Better Call GPT, Comparing Large Language Models Against Lawyers. *Working Paper.*
https://arxiv.org/html/2401.16212v1

Learnprompting. (2024). *Role Prompting doesn't work....* July 15, 2024

Lebovitz, S., Levina, N., & Lifshitz-Assaf, H. (2021). Is AI ground truth really true? The dangers of training and evaluating ai tools based on experts' know-what. *MIS Quarterly, 45*(3), 1501-1525.

Lebovitz, S., Lifshitz-Assaf, H., & Levina, N. (2022). To engage or not to engage with AI for critical judgments: How professionals deal with opacity when using AI for medical diagnosis. *Organization Science, 33*(1), 126-148.

Lee, G. (2024). *AI could increase growth by 1.5% over the next 10 years, Goldman Sachs says.* CNBC. February 14, 2024 https://www.cnbc.com/video/2024/02/14/ai-could-increase-growth-by-1point5percent-over-the-next-10-years-goldman-sachs.html

Lee, P., Goldberg, C., & Kohane, I. (2023). *The AI Revolution in Medicine: GPT-4 and Beyond.* Pearson.

Lee, T. (2023). *In retrospect, the tech hype is wrong.* April 24, 2023 https://x.com/binarybits/status/1650487310880743426

Lemkin, J. (2023). *Single most voracious demand for AI is contact center.* July 7, 2023 https://x.com/jasonlk/status/1677115929006854144

Leopold, A. (2024). *When it started becoming clear to some that an atomic bomb was possible, secrecy, too, was perhaps the most contentious issue.* June 7, 2024 https://x.com/leopoldasch/status/1798820621973111147

Levy, S. (2023). How Not to Be Stupid About AI, With Yann LeCun. *Wired Magazine.* https://www.wired.com/story/artificial-intelligence-meta-yann-lecun-interview/

Li, J., Chen, J., Ren, R., Cheng, X., Wayne Xin Zhao, Nie, J.-Y., & Wen, J.-R. (2024). *The Dawn After the Dark: An Empirical Study on Factuality Hallucination in Large Language Models.* Working Paper. https://arxiv.org/html/2401.03205v1

Li, J., Zhang, Q., Yu, Y., Fu, Q., & Ye, D. (2024). More Agents Is All You Need. *Working Paper.* https://arxiv.org/abs/2402.05120

Lindebaum, D., & Fleming, P. (2023). ChatGPT Undermines Human Reflexivity, Scientific Responsibility and Responsible Management Research. *British Journal of Management, 35*(2), 566-575.

Lior. (2023). *Just found out about the Chatbot Arena, it's brilliant. It allows you to compare and rank the output of 25+ LLMs right from your browser..* 2023-12-20 https://x.com/AlphaSignalAI/status/1737537992703844521

Lott, M. (2024). *Massive breakthrough in AI intelligence: OpenAI passes IQ 120.* September 14, 2024 https://www.maximumtruth.org/p/massive-breakthrough-in-ai-intelligence

Lysyakov, M., & Viswanathan, S. (2022). Threatened by AI: Analyzing Users' Responses to the Introduction of AI in a Crowd-Sourcing Platform. *Information Systems Research.*

https://pubsonline.informs.org/doi/abs/10.1287/isre.2022.1
184

Management, A. I. (2023). *Ark Investment 2023 Generative AI.* Ark
    Investment Management LLC. January 31, 2023 Ark
    Investment Management LLC

Marcus, G. (2024). *SCOOP: OpenAI may lose $5B this year & may
    run out of cash in 12 months, unless they raise more $.*
    2024-07-24
    https://twitter.com/GaryMarcus/status/1816116071226868o
    85

Martin, D. (2024). 6 Bold Statements By Nvidia CEO Jensen Huang
    On AI's Future. *CRN Channel Co.* August 5, 2024

Marwaha, A. (2017). *7 Ways artificial intelligence can benefit your
    law firm.* American Bar Association.
    https://www.americanbar.org/news/abanews/publications/y
    ouraba/2017/september-2017/7-ways-artificial-intelligence-
    can-benefit-your-law-firm/

Masad, A. (2024). *Announcing Replit Agent.* Replit. September 6,
    2024

McElheran, K., Li, J. F., Brynjolfsson, E., Kroff, Z., Dinlersoz, E.,
    Foster, L., & Zolas, N. (2024). AI adoption in America: Who,
    what, and where. *Journal of Economics & Management
    Strategy, 33*(2), 375-415.

Meglio, F. D. (2023). *What do HR Leaders really think about
    ChatGPT?* HR Exchange Network.
    https://www.hrexchangenetwork.com/hr-
    tech/articles/what-do-hr-leaders-really-think-about-chatgpt

Mencher, A. G. (1971). *On the Social Deployment of Science* (Vol. 27).
    Bulletin of the Atomic Scientists.

Miglio, A. D., Giovine, C., Hauser, S., Ouass, M., & Wildt, N. V. d.
    (2024). *Banking on innovation: How ING uses generative
    AI to put people first.* Mckinsey and Company.
    https://www.mckinsey.com/industries/financial-
    services/how-we-help-clients/banking-on-innovation-how-
    ing-uses-generative-ai-to-put-people-first

Mikhail, P. (2023). *Many people don't realize it: you can analyze
    even your local documents, the ones not available on the
    web.* . 2023-04-28
    https://twitter.com/MParakhin/status/165161865997310771
    3

Miller, A. K. (2023). *Enterprise: Business Strategy.* 2023-06-12
    https://twitter.com/alliekmiller/status/16682538966098739
    20

Miller, A. K. (2024a). *AI experts refer to prompt engineering.* 2024-07-22
https://x.com/alliekmiller/status/1815063763068088757

Miller, A. K. (2024b). *Perplexity's new Perplexity Pages feature.* 2024-07-01
https://x.com/alliekmiller/status/1796618758712152425

Mollick, E. (2023a). *The Homework Apocalypse.* July 1, 2023
https://www.oneusefulthing.org/p/the-homework-apocalypse

Mollick, E. (2023b). *If AI really does plateau at 60-80th percentile of human ability (no sign it will/won't), the impacts may be stabilizing.* . 2023-11-27
https://x.com/emollick/status/1784359024592359587

Mollick, E. (2023c). *Prediction Markets and AI labs suggest AGI is within planning horizons.* November 28, 2024
https://x.com/emollick/status/1729295140840202481

Mollick, E., & Mollick, L. (2023). Assigning AI: Seven Approaches for Students, with Prompts
[http://dx.doi.org/10.2139/ssrn.4475995]. *Working Paper.*

Mukherjee, A., & Chang, H. H. (2023). Managing the Creative Frontier of Generative AI: The Novelty-Usefulness Tradeoff. *California Management Review Insights.*

Murad. (2024). *AI and Automation is going to be replacing more & more jobs every year (Stanford Study).* 2024-02-22
https://x.com/MustStopMurad/status/1787795402168623130/

Musk, E. (2018). Elon Musk Podcast Transcript, *Joe Rogan Experience.* September 7, 2018

Musk, E. (2020). Elon Musk Podcast Transcript, *Joe Rogan Experience.* May 7, 2020

Narayanan, A., & Kapoor, S. (2024). *GPT-4 and Professional Benchmarks.* AI Snake Oil.
https://www.aisnakeoil.com/p/gpt-4-and-professional-benchmarks

Naumovska, I., Gaba, V., & Greve, H. (2021). The Diffusion of Differences: A Review and Reorientation of 20 Years of Diffusion Research. *Academy of Management Annals, 15,* 377-405.

Naveed, H., Khan, A. U., Qiu, S., Saqib, M., Anwar, S., & Mian, A. (2023). A comprehensive overview of large language models. *Working Paper.* https://arxiv.org/abs/2307.06435

Neff, A. (2024). How Will Generative and Conversational AI Impact Call Centers? March 19, 2024 https://www.icmi.com/resources/2024/how-generative-and-conversational-ai-will-impact-contact-centers

Ng, A. (2023). *Gptfile, a way to organize files with natural language using gpt-4.* May 30, 2023 https://x.com/localghost/status/1663274587860393984

Ng, A., & Fulford, I. (2023). *ChatGPT Prompt Engineering for Developers.* April 27, 2023 https://x.com/AndrewYNg/status/1651605660382134274

O'Donnell, J. (2024). Sam Altman says helpful agents are poised to become AI's killer function. *MIT Technology Review.* May 1, 2024 https://www.technologyreview.com/2024/05/01/1091979/sam-altman-says-helpful-agents-are-poised-to-become-ais-killer-function/

Obasanjo, D., & Zitron, E. (2024). *It is disconcerting that every piece of software you use for work today whether it's Asana, GitHub or Zoom is one update away from their sales team pitching your employer that you can be replaced by it based on being trained by your usage of the product.* . 2024-07-09 https://x.com/edzitron/status/1810362077867028497

Olenick, M., & Zemsky, P. (2023). Can GenAI do strategy? *Harvard Business Review.* https://hbr.org/2023/11/can-genai-do-strategy

OpenAi. (2023). *GPT-4 Technical Report.* https://cdn.openai.com/papers/gpt-4.pdf

OpenAI. (2024). *Learning to Reason with LLMs.* September 12, 2024 https://openai.com/index/learning-to-reason-with-llms/

Otis, N. G., Clarke, R., Delecourt, S., Holtz, D., & Koning, R. (2023). The uneven impact of generative AI on entrepreneurial performance. *Working Paper.* https://www.hbs.edu/ris/Publication%20Files/24-042_9ebd2f26-e292-404c-b858-3e883f0e11c0.pdf

Paik, C. (2024). *The End of Software.* June 1, 2024 https://x.com/cpaik/status/1796633683908005988

Palazzolo, S., & Efarti, A. (2024). OpenAI Shifts AI Battleground to Software that Operates Devices, Automates Tasks. *The Information.* https://www.theinformation.com/articles/openai-shifts-ai-battleground-to-software-that-operates-devices-automates-tasks?utm_source=ti_app

Palazzolo, S., Woo, E., & Efrati, A. (2024). OpenAI Shifts Strategy as Rate of 'GPT' AI Improvements Slows. *The Information.*

https://www.theinformation.com/articles/openai-shifts-strategy-as-rate-of-gpt-ai-improvements-slows

Palihapitiya, C. (2023a). *Software engineers completed a coding task in less than half the time with AI coding assistant GitHub Copilot.* August 5, 2023 https://x.com/chamath/status/1687568795865317376

Palihapitiya, C. (2023b). *US researchers showed that within a few months of the launch of ChatGPT, copywriters and graphic designers on major online freelancing platforms saw a significant drop in the number of jobs they got, and even steeper declines in earnings.* November 10, 2023 https://x.com/chamath/status/1722855387123359786

Patel, D. (2024). *Devin got wrecked in 3 weeks by the open source AutoCodeRover.* April 9, 2024 https://x.com/dylan522p/status/1777623660829769781

Patel, K. (2024). *Gartner Hype Cycle for Artificial Intelligence, 2023.* May 23, 2024 https://x.com/KrisPatel99/status/1793382809463144841

Peng, S., Kalliamvakou, E., Cihon, P., & Demirer, M. (2023). The Impact of AI on Developer Productivity: Evidence from Github Copilot. *Working Paper.* https://arxiv.org/abs/2302.06590

Peter, W. (2024). *What I'm expecting in the next 5-6 years of AI development.* 2024-08-20 https://x.com/peterwildeford/status/1825614599623782490

Philipp Koralus, V. W.-M. (2023). Humans in Humans Out: On GPT Converging Toward Common Sense in both Success and Failure. *Working Paper.* https://arxiv.org/abs/2303.17276

Pi. (2024). *Pi.ai.* https://pi.ai/discover

Plumb, T. (2024). AI pioneer LeCun to next-gen AI builders: 'Don't focus on LLMs'. *VentureBeat.* May 22, 2024 https://venturebeat.com/ai/ai-pioneer-lecun-to-next-gen-ai-builders-dont-focus-on-llms/

Puranam, P. (2021). Human–AI collaborative decision-making as an organization design problem. *Journal of Organization Design, 10*(2), 75-80.

Rachel, W. (2024). *My recommendation to "learn AI" has become much more nuanced after teaching hundreds of people over the past 18 months.* 2024-06-26 https://twitter.com/rachel_1_woods/status/1805811212493496804

Raisch, S., & Fomina, K. (2024). Combining human and artificial intelligence: Hybrid problem-solving in organizations.

Academy of Management Review.
https://doi.org/10.5465/amr.2021.0421

Raisch, S., & Krakowski, S. (2021). Artificial intelligence and management: The automation–augmentation paradox. *Academy of Management Review, 46*(1), 192-210.

Rathi, I., Tayler, S., Bergen, B. K., & Jones, C. R. (2024). GPT-4 is judged more human than humans in displaced and inverted Turing tests. *Working Paper.*
https://arxiv.org/pdf/2407.08853

Reddy, B. (2024). *The Power of Prompt Engineering.* July 20, 2024
https://x.com/bindureddy/status/1814409737557160044

Robert Irvine, D. B., Vyas Raina, Adian Liusie, Ziyi Zhu, Vineet Mudupalli, Aliaksei Korshuk, Zongyi Liu, Fritz Cremer, Valentin Assassi, Christie-Carol Beauchamp, Xiaoding Lu, Thomas Rialan, William Beauchamp. (2023). Rewarding Chatbots for Real-World Engagement with Millions of Users. *Working Paper.* https://arxiv.org/abs/2303.06135

Robins, J. (2023). *Meet the generative AI startups pulling in the most cash.* Pitchbook. October 18, 2023
https://pitchbook.com/news/articles/vc-valuations-generative-ai-startups

Roemmele, B. (2023a). *DoctorGPT is an LLM that can pass the US Medical Licensing Exam.* August 14, 2023
https://x.com/BrianRoemmele/status/1691076100498276352

Roemmele, B. (2023b). *Open source local models on the path to overtake massive (and expensive) cloud based closed models.* December 12, 2023
https://x.com/BrianRoemmele/status/1734333713381753165?s=20

Roemmele, B. (2024). *Powerful summaries from YouTube videos.* Feb 2, 2024
https://x.com/BrianRoemmele/status/1753111424514371906

Roon. (2024). *https://x.com/tszzl.* https://x.com/tszzl

Roth, E. (2024). ChatGPT's weekly users have doubled in less than a year / Now 200 million people use the AI chatbot each week. *The Verge.* April 30, 2024
https://www.theverge.com/2024/8/29/24231685/openai-chatgpt-200-million-weekly-users

Rotman, D. (2023). ChatGPT is about to revolutionize the economy. We need to decide what that looks like. *MIT Technology Review.* March 25, 2023
https://www.technologyreview.com/2023/03/25/1070275/chatgpt-revolutionize-economy-decide-what-looks-like/

Salesforce. (2023). *Introducing the ChatGPT App for Slack.* Salesforce. 2023-03-22 https://www.salesforce.com/news/stories/chatgpt-app-for-slack/

Sato, M. K., Koba, L. J., Du, H., Goodrich, B., Hasin, M., Chan, L., Miles, L. H., Lin, T. R., Wijk, H., Burget, J., Ho, A., Barnes, E., & Christiano, P. (2023). Evaluating Language-Model Agents on Realistic Autonomous Tasks. In.

Schulhoff, S., Ilie, M., Balepur, N., Kahadze, K., Liu, A., Si, C., Li, Y., Gupta, A., Han, H., Schulhoff, S., Dulepet, P. S., Vidyadhara, S., Ki, D., Agrawal, S., Pham, C., Kroiz, G., Li, F., Tao, H., Srivastava, A., . . . Resnik, P. (2024). The Prompt Report: A Systematic Survey of Prompting Techniques. *Working Paper.* https://arxiv.org/pdf/2406.06608

Schulman, A. (2009). Peter Thiel on the Singularity and economic growth. *The New Atlantis.* October 4, 2009 https://www.thenewatlantis.com/futurisms/peter-thiel-on-singularity-and-economic

Sharyph. (2023). *ChatGPT Prompts to Analyze Data.* May 7, 2023 https://x.com/sharyph_/status/1655106890534141952?s=20

Shervin Minaee, T. M., Narjes Nikzad, Meysam Chenaghlu Richard Socher, Xavier Amatriain, Jianfeng Gao. (2024). Large Language Models: A Survey. *Working Paper.* https://arxiv.org/pdf/2402.06196

Sheth, A. (2023). *Twitter Post In My Style: Mimics your specific formatting and writing style.* May 4, 2023 https://x.com/aaditsh/status/1654139807704940544

Sheth, A. (2024). *AI-Powered Hiring Made Easy.* Prompts Daily. https://www.neatprompts.com/p/ai-hiring-the-perfect-candidate

Shipper, D. (2023). *How to build a chatbot in GPT-4.* April 18, 2023 https://x.com/danshipper/status/1648313206429949953

Shyan, L. S. (2024, January 21, 2024). Is the tech boom tapering off? *The Straits Times.* https://www.straitstimes.com/business/invest/is-the-tech-boom-tapering-off

Snowden, R. (2023). *How Can Human Resources Utilise ChatGPT?* Jan 27, 2023 https://medium.com/@rsnowden21/how-can-human-resources-utilise-chatgpt-fe8e74ec4e6a

Solow, R. (1987). We'd better watch out. *New York Times Review of Books.*

Song, F., Agarwal, A., & Wen, W. (2024). The Impact of Generative AI on Collaborative Open-Source Software Development:

Evidence from GitHub Copilot. *Working Paper.*
https://www.researchgate.net/publication/384630465_The_Impact_of_Generative_AI_on_Collaborative_Open-Source_Software_Development_Evidence_from_GitHub_Copilot

Srinivas, A. (2024). *The trend is clear. Bet your money on small open-source models, distillation and fine-tuning, serving, and data collection. .* 2024-07-25
https://x.com/AravSrinivas/status/1816248208802336975

Staff, A. (2023). *7 Inspirational Demis Hassabis Quotes on AI's Future.* August 17, 2023 https://www.aiifi.ai/post/demis-hassabis-quotes

Stelzer, F. (2023). *New "productivity paradox" just dropped.* 2023-07-22
https://twitter.com/fabianstelzer/status/1682437035204681760

Stokes, C. (2018). Why the three laws of robotics do not work. *International Journal of Research in Engineering and Innovation,* 2(2), 121-126.

Storti, L. (2023). *ChatGPT combined with modern marketing is going to make people millions in 2023.* 2023-01-03
https://twitter.com/loganstorti/status/1610251544846565376?s=20

Tafti, E. A. (2023). Technology, Skills, and Performance: The Case of Robots in Surgery. *Working Paper.*
https://ideas.repec.org/p/ajt/wcinch/78746.html

Tao, T. (2024). Embracing change and resetting expectations. *AI Anthology.* https://unlocked.microsoft.com/ai-anthology/terence-tao/

Tom B. Brown, B. M., Nick Ryder, Melanie Subbiah, Jared Kaplan, Prafulla Dhariwal, Arvind Neelakantan, Pranav Shyam, Girish Sastry, Amanda Askell, Sandhini Agarwal, Ariel Herbert-Voss, Gretchen Krueger, Tom Henighan, Rewon Child, Aditya Ramesh, Daniel M. Ziegler, Jeffrey Wu, Clemens Winter, Christopher Hesse, Mark Chen, Eric Sigler, Mateusz Litwin, Scott Gray, Benjamin Chess, Jack Clark, Christopher Berner, Sam McCandlish, Alec Radford, Ilya Sutskever, Dario Amodei. (2020). Language Models are Few-Shot Learners. *Working Paper.*
https://arxiv.org/abs/2005.14165

Toor, H. (2024). *An AI-powered meeting recorder.* June 26, 2023

Tossell, B. (2024a). *How Clearbit uses AI - They were acquired for $150M by Hubspot last year.* January 25, 2024
https://x.com/bentossell/status/1750492229288779838

Tossell, B. (2024b). *Insights from Stripe, Google, Intercom, Zapier, Adobe and others and came up with 9 ways they help encourage employee adoption.* Februrary 1, 2024 https://x.com/bentossell/status/1753040606526394572

Tripsas, M. (2009). Technology, Identity, and Inertia Through the Lens of "The Digital Photography Company. *Organization Science, 20,* 441-460.

Tsarathustra. (2024). *Richard Sutton says AI safety advocates are creating the opposite of what they seek.* August 11, 2024 https://x.com/tsarnick/status/1822406616320454953

Tully, T., Redfem, J., & Xiao, D. (2024). *The State of Generative AI in the Enterprise.* Menlo Ventures. **https://menlovc.com/2024-the-state-of-generative-ai-in-the-enterprise/**

Tunguz, B. (2024). *Recruiters spend 3-to-5 seconds on a resume.* January 26, 2024 https://x.com/tunguz/status/1750658222204002579?s=46

Turck, M. (2024). *Full Steam Ahead: The 2024 MAD (Machine Learning, AI & Data) Landscape.* . https://mattturck.com/mad2024/

Tushman, M. L., & Anderson, P. (1986). Technological Discontinuities and Organizational Environments. *Administrative Science Quarterly, 31,* 439-465.

Tyler Hutcherson, H. C. (2023). Build an E-commerce Chatbot With Redis, LangChain, and OpenAI. https://redis.com/blog/build-ecommerce-chatbot-with-redis/

Vanneste, B. S., & Puranam, P. (2024). Artificial Intelligence, trust, and perceptions of agency. *Academy of Management Review.* ttps://doi.org/10.5465/amr.2022.0041

Verma, P., & Vynck, G. D. (2023, June 2, 2023). ChatGPT took their jobs. Now they walk dogs and fix air conditioners. *The Washington Post.* https://www.washingtonpost.com/technology/2023/06/02/ai-taking-jobs/

Wei, J., Tay, Y., Bommasani, R., Raffel, C., Zoph, B., Borgeaud, S., Yogatama, D., Bosma, M., Zhou, D., Metzler, D., Chi, E. H., Hashimoto, T., Vinyals, O., Liang, P., Dean, J., & Fedus, W. (2022). Emergent abilities of large language models. *Working Paper.* https://arxiv.org/abs/2206.07682

Weng. (2023). *Companies are hiring like crazy for Generative AI talent.* September 11, 2023 https://x.com/AznWeng/status/1701228289308721316

Willison, S. (2024). *Datasette Extract is a new Datasette plugin that uses GPT-4 (and the new GPT-4 Vision) to extract structured data from unstructured text and images and insert it into a SQLite database table.* April 10, 2024 https://x.com/simonw/status/1777820654487925245

Winton, B. (2023). *The future is faster than you think.* November 25, 2023 https://x.com/wintonARK/status/1728139361903075744

Wolfe, C. R. (2024). *RAG is one of the best (and easiest) ways to specialize an LLM over your own data.* February 6, 2024 https://x.com/cwolferesearch/status/1754558231802769857

Woodside, T. (2023). *Updated graph of NVIDIA datacenter revenue, with the latest quarter.* . 2023-11-25 https://twitter.com/Thomas_Woodside/status/17281953393 31751955

Wu, S. (2024). *Introducing Devin, the first AI software engineer.* Cognition Labs. March 12, 2024

Wu, S., İrsoy, O., Lu, S., Dabravolski, V., Dredze, M., Gehrmann, S., Kambadur, P., Rosenberg, D., & Mann, G. (2023). BloombergGPT: A Large Language Model for Finance. *Working Paper.* https://arxiv.org/html/2303.17564v3

Xiang, H., Reshef, O., & Luofeng, Z. (2023). The Short-Term Effects of Generative Artificial Intelligence on Employment: Evidence from an Online Labor Market. *Working Paper.* http://dx.doi.org/10.2139/ssrn.4527336

Xianzhi Li, S. C., Xiaodan Zhu, Yulong Pei, Zhiqiang Ma, Xiaomo Liu, Sameena Shah. (2023). Are ChatGPT and GPT-4 General-Purpose Solvers for Financial Text Analytics? A Study on Several Typical Tasks. *Working Paper.* https://arxiv.org/abs/2305.05862

Yohei. (2024). *List of technologies, jobs it killed, and jobs it created.* January 26, 2024 https://x.com/yoheinakajima/status/1750653097129758889

Zakin, P. (2024). *The biggest consumer internet companies have been started in the wake of technological unlocks.* 2024-08-26 https://x.com/pzakin/status/1827779647725318341

Ziniti, C. (2024). *Stanford's working on a study showing most AIs fail at legal work.* . June 6, 2024

Zitron, E. (2024). *Goldman Sachs has called BS on Generative AI.* July 8, 2024 https://www.wheresyoured.at/pop-culture/

Ziwei Ji, N. L., Rita Frieske, Tiezheng Yu, Dan Su, Yan Xu, Etsuko Ishii, Yejin Bang, Delong Chen, Wenliang Dai, Ho Shu Chan, Andrea Madotto, Pascale Fung. (2022). *Survey of*

Hallucination in Natural Language Generation. Working Paper. https://arxiv.org/abs/2202.03629

Zoë Schiffer, C. N. (2023). *Amazon's Q has 'severe hallucinations' and leaks confidential data in public preview, employees warn.* Platformer. December 1, 2023 https://www.platformer.news/amazons-q-has-severe-hallucinations/

Zuboff, S. (1985). *In the Age of the Smart Machine: The Future of Work and Power.* Basic Books.